THE
LONG
THAW

Books in the **SCIENCE ESSENTIALS** series bring cutting-edge science to a general audience. The series provides the foundation for a better understanding of the scientific and technical advances changing our world. In each volume, a prominent scientist—chosen by an advisory board of National Academy of Sciences members—conveys in clear prose the fundamental knowledge underlying a rapidly evolving field of scientific endeavor.

DAVID ARCHER

THE
LONG
THAW

How Humans Are Changing the Next
100,000 Years of Earth's Climate

Princeton University Press

Princeton and Oxford

In the United Kingdom: Princeton University Press,
6 Oxford Street, Woodstock,
Oxfordshire OX20 1TW

press.princeton.edu

Fifth printing, and first paperback printing, 2010
Paperback ISBN: 978-0-691-14811-3
Cloth ISBN: 978-0-691-13654-7

Library of Congress Control Number: 2008931528

British Library Cataloging-in-Publication Data
is available

This book has been composed in Minion
and Impact

This book is printed on recycled paper ♻

Printed in the United States of America

9   10   8

For my parents and their friends

in the dinner / discussion group

Contents

Acknowledgments

This manuscript benefited from comments by Jeffrey Kiehl, Ingrid Gnerlich, Daniel Yang, and an anonymous reviewer, and by conversations with Ken Caldeira and Andy Ridgwell. I gave many public presentations of these ideas in the course of writing the book, and I was asked a lot of insightful questions. I also get tons of feedback from the readers of the climate science Web site realclimate.org. If you are one of the many people I've interacted with on these topics, in whichever venue, please accept my thanks.

Global Warming
in Geologic Time

Global warming could be one of humankind's longest lasting legacies. The climatic impacts of releasing fossil fuel CO_2 to the atmosphere will last longer than Stonehenge. Longer than time capsules, longer than nuclear waste, far longer than the age of human civilization so far. Each ton of coal that we burn leaves CO_2 gas in the atmosphere. The CO_2 coming from a quarter of that ton will still be affecting the climate one thousand years from now, at the start of the next millennium. And that is only the beginning.

The excess CO_2 in the atmosphere at the next millennium may not be the exact same molecules that came from our power plants. Some of the CO_2 from fossil fuels will have been taken up into trees, or deposited in soils. Some will have dissolved in the oceans. But, as this book will explain, the CO_2 concentration in the atmosphere at the next millennium will be higher if that coal is burned than if it is not. About 10% of the CO_2 from coal will still be affecting the climate in one hundred thousand years.

Over the last few centuries, mankind has been humbled by insights from the scientific enterprise. Darwin told us that humans are not biologically "special"; we are descended from monkeys, and they from even humbler origins. Copernicus discovered that the Earth is not the center of the universe, but rather revolves around the Sun, an ordinary star like billions of others. Geologists' reconstruction of the history of the Earth tells us that the world is much older than we are, and there's no evidence that it was created especially for us. Most of Earth history predates the arrival of humans. This is all very humbling. Global climate is a canvas upon which mankind may be painting one of his longest-lasting legacies. We're not so puny, after all. We are becoming players in geologic time.

The first section of the book is a snapshot of the situation we find ourselves in right now. In geologic time, a century is nothing, an eyeblink, so let's be geological and consider the last century and the next century to be "the present." The theory behind global warming has been around for about a century. Meanwhile, the CO_2 concentration in the atmosphere has been steadily rising. This was discovered about a half-century ago. Just in the last few decades, the temperature of the atmosphere has begun to rise in a way that can be satisfactorily explained only by the greenhouse theory, which has the implication that it will get even warmer if CO_2 continues to rise. The first section of the book contains an explanation of what the forecast calls for, and why.

The second section of the book is about the past. A tenet at the foundation of geology is that the present is the key to the past. The idea is that processes that can be observed today might also be responsible for things that happened in the past, given the vast stretches of geological time. An ice sheet grinds up the rocks into dust, which blows away and deposits someplace else. Eventually, after tens of thousands of years, you have a layer of glacial flour many meters thick.

In this book we are operating on a somewhat different philosophical premise, using the past as the key to the future. Our motivating interest here is the forecast for global warming. How bad does it sound? How likely is it? Is global warming something new, or is this something that happens all the time?

Global warming is not the first climate event in Earth history. There were even larger climate changes in the past. There were sudden climate flip-flops, a switch in a few years from one climate to another that lasted a thousand years. There was also the slow, ponderous transition from the tropical world of the dinosaurs to the icy world of today.

Reconstructed climate changes of the past can be used to test the models that are used to forecast the future. Climatology is not an experimental science, in that we don't do experiments with the real climate system, at least not intentionally. So one approach to understanding the climate system is to reconstruct how it varied in the past, and how it responded to getting poked in various ways. A grant proposal would probably describe the paleoclimate record as a "natural laboratory." A graduate student would probably say that we are using the past to tune our chops for forecasting the future.

Climate changes of the past also help visualize and calibrate the forecast for the future. The global average temperature of the Earth might be 3 °C warmer in the year 2100 than it was in 1950. This doesn't sound like much; as I look outside it is probably at least 3 °C warmer at this moment than it was early this morning, and the world doesn't seem to be ending. On the other hand, the climate changes that civilized humanity has witnessed have all been 1°C or less. Earth has warmed almost this much already because of human activity, but this is nothing compared with the forecast for 2100.

The second section of this book describes climate changes in the past, as they are relevant to the global warming forecast for the future. The first three chapters (4, 5, and 6) describe three different modes of natural climate variability, which operate on three very different timescales. Chapter 7 brings together the present and the past. The impatient reader could even skip ahead to Chapter 7 without losing the thread of the argument too much. Think of Chapter 7 as a sort of Executive Summary So Far.

Against this backdrop, in Section 3, we turn our attention to the deep future of the global warming climate event. The excess CO_2 in the atmosphere is absorbed and transformed into carbon in trees, mucky soils, and dissolved in the ocean, and so the warming begins to subside. The oceans are a big player in this story, absorbing a majority of the CO_2 we release, on a timescale of a few centuries.

Early earth scientists believed, perhaps without giving it too much thought, that CO_2 would invade the ocean more quickly than this, perhaps fast enough that the CO_2 concentration of the atmosphere would be essentially imperturbable by human activity. This is a blue planet, after all. The ocean covers three quarters of the the surface of the Earth.

But most of the water in the ocean is cold, deep abyssal water, which sees the atmosphere only maybe every thousand years. The pathway for fossil fuel CO_2 into the deep ocean is through the surface ocean in very cold places like near Antarctica or Greenland, which cover only a small fraction of the Earth's surface. This bottleneck is the reason for the centuries it will take for fossil fuel CO_2 to dissolve in the oceans.

Several centuries seems—to me, personally—like a pretty long time. Mozart lived several centuries ago. But for human-induced climate change, this is only the beginning. It turns out that, after a new slug of CO_2 has spread out in whatever way it chooses between the atmosphere and the ocean, there will still be excess CO_2 left in the atmosphere.

The ultimate fate of that leftover CO_2 will be to react with dissolving rocks. Chemically, igneous rock acts like a base, able to neutralize and thereby absorb CO_2, which is an acid. Eventually, the carbon winds up as limestone shells on the ocean floor. Carbon emerges from the solid Earth during fossil fuel combustion, and it returns to the solid Earth as limestone.

The kicker is that it will take thousands of years, even hundreds of thousands of years, for these chemical reactions with rocks to completely scrub the planet of our extra CO_2. Most of the excess CO_2 in the atmosphere goes away in a few centuries by dissolving in the ocean, but the rest has to wait. The atmospheric CO_2 peak has a long tail (Figure 14 in Chapter 8).

Mankind has a kind of vested interest in time spans of centuries. I personally can visualize centuries. I like to think that Benjamin Franklin's childhood was not unimaginably different from my own. (OK, so I probably watched more TV than he did.) I know people who knew people who knew the beginning of the last century. I can look the last century in the eye.

Looking forward, a century is about how far I can really imagine also. Sixty years is grandchildren. One hundred is great grandchildren or great, great grandchildren. After that, they're on their own, am I right?

Climate change science and politics are also very focused on the century time horizon. There are records of temperatures, measured by thermometers, going back about a century. The Intergovernmental Panel on Climate Change (IPCC) Scientific Assessment forecast of global warming, described in Chapter 1, is particularly focused on climate change between now and the year 2100.

It makes perfect sense to focus our attention on a time span that is imaginable given our own century-timescale lifetimes. Even if we lived to be a thousand years old, or ten thousand, the century timescale would be an important one to watch, no doubt about it. One reason is that fossil fuel CO_2 is released on a timescale of centuries. By the year 2100, traditional oil and gas will be gone. It might take a few centuries to burn all the coal, however, and the coal is where most of the carbon is. Atmospheric CO_2 goes up and then comes most of the way back down, as it is absorbed in the oceans, within a few centuries. Therefore the largest climate changes will consist of a climate storm, several centuries long, significantly worse than the forecasts for the year 2100. Eventually, the storm will fade, mostly, into the long tail.

Much of the action in the global warming forecast takes place on timescales of centuries. When it comes time to make practical decisions about avoiding human-induced climate change, the century timescale is the first one to watch, no question.

But just out of curiosity (if for no other reason), let's consider climate changes on timescales that are much longer than that. Earth is more than a few centuries old. Human civilization is unique in Earth history but we are not the first climate provocateurs this old Earth has seen. Climate changes in the past can tell us a lot about the deep future. They have had time to play out, unlike our own global warming climate change, which is just beginning.

The deepest and most profound climate changes in the recent geologic past seem to take place on timescales of millennia and longer. The great ice sheets grow and usually melt on timescales of millennia, a huge response to the wobbles in the Earth's orbit. The natural carbon cycle acted as a positive feedback, amplifying the response to the orbit.

The climate of the Earth is so dramatically sensitive to the ten-thousand-year orbital wobbles that it might also put on a pretty good show in response to the long tail of the fossil fuel CO_2. I will try to convince you that human climate forcing has the potential to overwhelm the orbital climate forcing, taking control of the ice ages. Mankind is becoming a force in climate comparable to the orbital variations that drive the glacial cycles.

The Earth today is colder than the average over geologic time. Most of Earth's history has been ice-free. Over millions of years, the climate of the Earth drifts back and forth between an icy climate such as we have today, and a "hothouse" climate state. Forty million years ago, the Earth was in a hothouse climate called the Eocene Optimum. The climate was tropical to the poles, driven by atmospheric CO_2 concentration maybe 10 or 20 times higher than today.

The slow progression between hothouse and icy climates is driven by the cycling of CO_2 into and out of the solid Earth. CO_2 is released from the Earth in volcanic gases and hot springs at the bottom of the ocean. CO_2 is taken up by weathering reactions, the same reactions that will generate the long tail of the fossil fuel CO_2.

On a timescale of a million years or longer, the climate of the Earth is determined by the solid Earth, breathing in and out CO_2. The difference between an Eocene hothouse and an icy climate such as ours is determined by factors that affect carbon release or uptake, such as the arrangement of the continents, the uplift of mountains, the evolution of plants, and doubtless many other factors.

Volcanoes release much less CO_2 every year than we do, so the near-term future is going to be dominated by us. Human industry has taken its place alongside the natural climate forcing agents, with the distinction that we push things around 100 times faster than the natural ones typically do. Ultimately, the amount of fossil fuel available could be enough to raise the atmospheric CO_2 concentration higher than it has been in millions of years.

The sea level rise forecast for the coming century is 0.2 to about 0.6 meters. This forecast includes the effects of water expanding as it warms up and the water from melting mountain glaciers in places like Alaska. The forecast explicitly does not include what will ultimately be the most important process—the melting of major ice sheets in Greenland and Antarctica. Sea level changes in the past were perhaps one hundred times larger for a given warming than the IPCC forecast for the coming century. The large variations in the past were driven by growing and melting ice sheets.

IPCC refers to this possibility as "future rapid dynamical changes in ice flow," and concludes that ice sheet collapses are unlikely in the coming century, but impossible to predict reliably. The current state-of-the art computer models of ice sheets predict that the ice sheets will not melt into the ocean very much in the next hundred years. However, there are examples from the past of ice sheets collapsing into the ocean over only a few centuries. There are also rumblings in the ice today that suggest that ice sheets may know a few tricks about melting that the ice sheet models have still to figure out.

During the melting of the ice sheets, about fourteen thousand years ago, there was a time interval called Meltwater Pulse 1A during which the equivalent of three Greenland ice sheets melted into the ocean in just a few centuries. Sediments from the North Atlantic tell us of times called Heinrich events, between thirty and seventy thousand years ago, when the Laurentide ice sheet on North America collapsed within a few centuries, releasing "armadas of icebergs" into the Atlantic, floating as far south as Spain. If the Greenland ice sheet began to collapse into the ocean like this, it would be unstoppable, a century-long train wreck.

The ice today is rumbling. Seismometers on the ice feel more ice-quakes than they used to. Models of the Greenland ice sheet take centuries to respond to changes in climate, but the flow rates of the real Greenland ice sheet are accelerating already. The real ice sheets are far more sensitive to climate than the ice sheet models are. This is no evil conspiracy, but just unfinished business. Computational glaciology is a field on the move.

But on timescales of millennia and longer, no fancy new melting tricks are even required to drive alarming sea level changes.
Ice sheet models do not predict Meltwater Pulse 1A or the Heinrich events, but they do predict the eventual melting of Greenland if the local summertime temperature were 3°C warmer.
Greenland by itself would raise sea level by 7 meters if it melted.
   Sea level in the geologic past was much more responsive to
changes in global climate than what IPCC predicts for the year
2100. Past sea level varied by 10–20 meters for each 1°C change
in the global average temperature. The IPCC business-as-usual
forecast for 3°C would translate to 20–50 meters of sea level rise.
The changes in the map of the Earth would be obvious even
from space. It may take thousands of years for sea level to change
this much, but the long tail of the fossil fuel CO_2 gives us all the
time we need.

Why should we mere mortals care about altering climate 100,000
years from now? Climate change is forecast to the year 2100, a
date that very few people now reading this book will see, but a
time span considerably shorter than 100,000 years.
   The rules of economics, which govern much of our behavior,
tend to limit our focus to even shorter time frames. Values are
related across time using interest rates. A $100 obligation in 100
years might be dealt with by investing $5 today, at an inflation-
adjusted interest rate of 3% per year. A $100 cost in 500 years
shrinks to 0.003 cents today. Analyzed within the framework of
economics, a climate impact 100,000 years from now becomes
laughably irrelevant to any rational decision-making. I'm imagining the financial guys on television laughing smugly at how
irrelevant it is, ha ha ha.
   And yet, human cultural memory begins to approach the longevity of our proposed climate adventures. The earliest written
records date to 5500 years ago, and oral tradition may carry some
grain of history from earlier still. How would it feel if the ancient

Greeks, for example, had taken advantage of some lucrative business opportunity for a few centuries, aware of potential costs—such as, say, a stormier world, or the loss of 10% of agricultural production to rising sea levels—that could persist to this day? This is not how I want to be remembered.

Ancient economists might have assured us (as Bjørn Lomborg does today) that the proper course should be to skimp on clean-up, and invest the money instead. Get rich and deal with it. An investment made by the ancient Greeks will have blossomed into a real chunk of change by now, more than enough to pay us off for any damages inflicted. That sounds like a lot of trickle-down, to me.

It could be that future civilizations might simply adjust to the new climate regime. If humans had evolved in the Eocene, no doubt we would find it comfortable. But there are ways in which a hothouse world might be a real trade-down. We did not evolve in the Eocene, and our descendents might find it uncomfortable, just as we would. Large regions in continental interiors could dry out. Hurricanes could get stronger. Sea level rise could inundate 10% of the carrying capacity of the planet, or more. In the long run, it could be a steep price to pay for a century or so of fossil fuel energy.

If climate change turns out to be a disaster, there have been proposals to geo-engineer a return to a cooler climate. Large volcanic eruptions inject a haze of droplets and particles into the stratosphere, which reflects sunlight and measurably cools the climate for several years. We could put additives into commercial jet fuel that would produce stratospheric particles deliberately.

This and most other proposals for climate geo-engineering require an ongoing effort. The particles settle out a few years after the planes stop flying. If society some centuries down the line failed to pay its climate bill (a bill that we left to them, thank you very much), all of our accumulated CO_2 emissions would begin to impact climate within just a few years. Geo-engineering solutions seem rather puny next to the hundred-millennia lifetime of global warming.

The only geo-engineering scheme that deals with the persistence of CO_2 is to extract CO_2 from the atmosphere, to really clean it up. The CO_2 could be buried in the Earth, after reacting it artificially with rocks perhaps. The problem is that once CO_2 is diluted by releasing it to the atmosphere, it takes energy and work to un-mix it. Releasing our fossil fuel CO_2 to the atmosphere now is a phenomenally stupid strategy, if the eventual plan is to clean it back up.

Persistence is a factor in many other environmental issues. Nuclear power creates waste that must be stored and guarded for 10,000 years. Everyone knows that. The pesticide DDT is not very dangerous to animals when it is first applied, but its persistence in the environment allows it to build up over time to toxic concentrations in birds and mammals. Current pesticides avoid this problem by degrading more quickly. Some, such as the organophosphates, are more immediately toxic to animals, but that was the necessary trade-off. Freons, used in refrigerators, were engineered to be inert, and as a result, after they are released they survive long enough to reach the stratosphere, where they catalyze ozone destruction. New Freon-substitute chemicals are engineered to break down more quickly.

The long lifetime of fossil fuel CO_2 creates a sense of fleeting folly about the use of fossil fuels as an energy source. Our fossil fuel deposits, 100 million years old, could be gone in a few centuries, leaving climate impacts that will last for hundreds of millennia. The lifetime of fossil fuel CO_2 in the atmosphere is a few centuries, plus 25% that lasts essentially forever. The next time you fill your tank, reflect upon this.

Section 1

THE PRESENT

CHAPTER 1
The Greenhouse Effect

The global warming forecast is not new, nor has it changed much over the last century. The basic physics of the greenhouse effect was described in 1827 by Jean Baptiste Joseph Fourier. Fourier was a mathematician in Bonaparte's army in Egypt. His name is best known for the Fourier transform, a mathematical technique for separating some complicated signal (such as the history of temperature through time, to choose an apropos example) into the sum of simple waves of different frequencies (such as the day/night cycle and the annual cycle), what we call calculating a spectrum.

Fourier's contribution to Earth science is the idea that gases in the atmosphere that absorb infrared radiation could eventually warm up the surface of the earth. He made the analogy of a greenhouse, but the actual name "greenhouse effect" came later.

The temperature of a planet is set by a natural thermostat, which balances the planet's energy budget. Energy comes in to the Earth as sunlight and leaves as infrared. The greenhouse effect of a gas changes the outgoing part of the budget, the infrared. All objects warmer than absolute zero shine in the infrared. A hot heating element glows red that we can see; the same object at room temperature glows in the infrared.

The rate of energy loss from an object as infrared radiation depends on the temperature of the object. According to the Stefan–Boltzmann relation, the object loses energy at a rate of σT^4, where σ is the Stefan–Boltzmann constant (just a number one can look up in a reference book) and T^4 is the temperature of the Earth in kelvins raised to the fourth power. When the object is hot, it sheds energy much more quickly than when it is cool.

The planet balances its energy budget by warming up or cooling down until the energy loss to space equals the energy gain from the Sun, as in the top panel of Figure 1. The thermostat is a by-product of the need to balance the energy budget. The idea is analogous to water running through a sink, as in Figure 1, bottom panel. The faucet is on, and water is falling into the sink. The drain is open at the bottom of the sink, and the higher the water level is in the sink, the faster it will drain.

When the faucet is initially turned on, water flows in faster than it flows out, and the water level in the sink rises. The sink fills until water is running down the drain as quickly as it is coming out of the faucet. If we start the sink experiment with too much water in the sink, water would drain faster than it filled until it reached that same balancing water level.

If we give our no-atmosphere planet the same energy input from sunlight that the Earth enjoys, it would have an average temperature of about 3°F or −16°C, sub-freezing temperatures around the world. Fourier's greenhouse effect is what's keeping Earth so much warmer than this poor cold naked planet.

Fourier's insight was to add a layer of atmosphere to the planet, which absorbs and emits infrared radiation (Figure 2). The Earth's surface receives energy from the Sun, as before, and it also receives energy from infrared radiation shining down from the atmosphere. The temperature of the Earth's surface rises to about 86°F or 30°C. That's a bit on the high side, but much closer to the real temperature of the Earth.

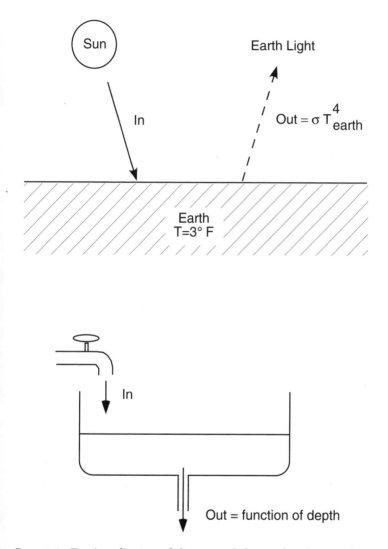

FIGURE 1. Top is a diagram of the energy balance of a planet with no atmosphere. The temperature of the planet finds the value at which energy outflow as infrared balances energy influx from the sun. The bottom is a sink, with water flowing in from a faucet and out down the drain. The rate of flow down the drain depends on the water level in the sink. The water level finds the value at which outflow balances inflow.

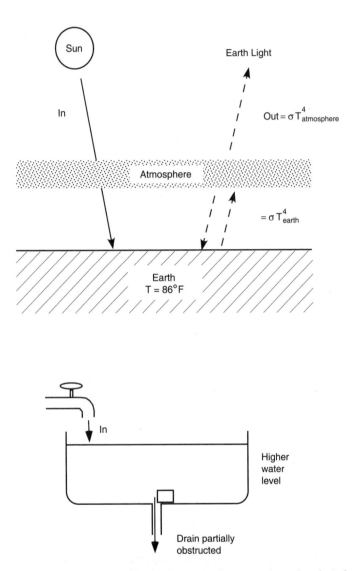

Figure 2. Top. A pane of glass analogous to the atmosphere absorbs infrared radiation from the ground, and radiates infrared at its own temperature. The atmosphere is colder than the ground, so the infrared radiation is impeded by the presence of the atmosphere. This is analogous to partially blocking the drain in the sink (bottom), which causes the water level in the sink to rise.

The greenhouse gas in Earth's energy balance is analogous to a partial obstruction of the drain at the bottom of the sink. A grape or piece of cucumber falls into the strainer, slowing down the drainage. The water level in the sink rises until it gets deep enough to force water through the obstructed drain as quickly as it flows in from the faucet. Let's hope the sink reaches a new balanced water budget before it overflows.

Just over a century ago, in 1896, Svante Arrhenius, a Swedish chemist, took the most astonishing leap I have ever read in climate science. Arrhenius used measurements of the brightness of infrared radiation from the moon to predict the temperature change you would get from raising CO_2. Arrhenius estimated a quantity which we now call the climate sensitivity, abbreviated as ΔT_{2x}. This is defined as the amount of warming that the Earth would undergo, on average, from a doubling of the atmospheric CO_2 concentration. The climate sensitivity is probably the first benchmark that two climate scientists in a bar would use to compare two different climate models.

The moonlight infrared data came from Samuel Pierpont Langley, who was trying to determine the temperature of the moon. The hotter the moon is, thought Langley, the brighter it shines in the infrared, the same story as for the Earth. "The dark rays" as they called them, were separated into different bands of wavelengths (different colors, if we could see them) by using a prism made of salt, because salt is one of the few solid substances that doesn't absorb infrared radiation. The intensity of the different invisible beams was measured using something called a bolometer, a device that measures the rate at which the invisible incoming light warms up a thermometer. It all must have seemed rather spooky.

Arrhenius used the data in a way that was not intended by Langley. Arrhenius looked for changes in the intensity of the "dark rays" that vary with humidity, and with the moon angle overhead, which affects the amount of atmosphere the light had to go through. In the moonlight data, more moonlight is absorbed when the light passes through more CO_2 or more water vapor. Arrhenius used this relationship in the data to predict how much the Earth would warm if you doubled CO_2. It was as though, analyzing the water flowing through our sink, Arrhenius calculated precisely how much the flow would slow down if you put a piece of carrot on the drain trap, obstructing the flow of water through the drain, and how much higher the water level in the sink would be.

The surface of the Earth does not all have the same temperature, though, the way that a sink has only one water level. Arrhenius did his calculation on a latitude and longitude grid, just as climate models do today, writing, "I should certainly not have undertaken these tedious calculations if an extraordinary interest had not been connected with them." After two years of pencil-and-paper arithmetic, he concluded that doubling the CO_2 concentration of the atmosphere would lead to 4 to 6°C of warming. Today, with the benefit of a century of innovation, hard work, and exploding computing power, we now estimate that doubling CO_2 would lead to about 2.5 to 4°C of warming. There have been revisions, discoveries, missteps, and wrong directions, as in any science, but on the whole not much has changed in the past century.

So what have climate scientists been doing in the meantime? Climate science has really exploded in the past few decades, as global warming grew from a prediction into an observation in the real world. Globally, about 2 billion dollars per year are being spent on climate change research, 50% of this in the United States. This sounds like a lot of money, and it is, but to put it into perspective, it amounts to only about 5% of the profits from the Exxon Mobil Oil Company. Much of the climate research

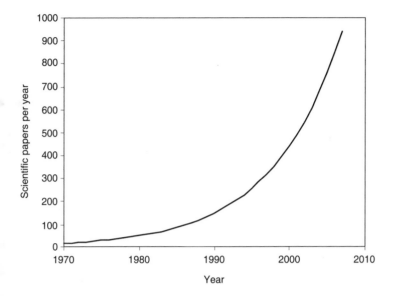

FIGURE 3. The rate of publication of scientific papers about climate in the past 35 years.

money is used to pay for satellites that monitor various aspects of the climate of the earth from space. Satellites are expensive. Meanwhile, thousands of scientists worldwide, at universities mostly, are hard at work developing climate models and theory, analyzing meteorological data, and reconstructing climates of the past. This form of research has an entrepreneurial feel to it: individuals or small groups, looking for the new angle that will get them funded and published. The scientific literature about global warming has exploded in the last decades, rising from about a hundred papers per year in the 1980s to a thousand per year today (Figure 3).

Climate science is interdisciplinary enough that it is a challenge to synthesize the bits and pieces. For example to understand climate change in the Arctic requires soil science, forestry, atmospheric and ocean physics, polar bear biology, and other scientific specialties. The state of the warming forecast for the entire globe encompasses so much information that no one human mind could hold it all at one time (not mine, anyway).

In response to warnings of the threat of global warming, the World Meteorological Organization created an organization of scientists charged with the task of summarizing the state of the science, called the Intergovernmental Panel on Climate Change or IPCC. The function of IPCC is not to do new research, but rather to summarize and synthesize all the published scientific papers into coherent reports. The scientists who do the actual work for IPCC are mostly employed by universities and national research laboratories around the world like NASA and NOAA. Working Group I of the IPCC writes the Scientific Assessment reports, while Working Group II reports on Impacts of climate change, and III on Mitigation (reducing CO_2 emissions, mostly). The most recent IPCC reports were released in spring of 2007. The projections and impacts of global warming as presented in the next two chapters are based on information from this report.

Most of the major ingredients in the global warming forecast were there in the results of Arrhenius' tedious calculations. One important example is called the ice albedo feedback. The word albedo describes the reflectivity of a planet to visible sunlight. Clouds reflect sunlight, as does ice and snow. When sunlight is reflected to space, it would be analogous to water from the faucet in the sink analogy that splashes onto the floor. Since that water doesn't have to go down the drain, the water level in the sink decreases. The water level is analogous to Earth's temperature, which falls if more incoming sunlight is reflected back to space instead of being absorbed.

The coupling between ice and light works out to be a loop of cause and effect called a feedback. The air warms for some external reason like rising CO_2, and as a result ice and snow melt on the land or ocean surface. The ice and snow are very reflective, which helped keep the planet cool, but the ground or ocean underneath have a greater tendency to absorb incoming sunlight, so the planet warms more than it would have. This is an example of a positive, amplifying feedback.

The Arctic warms more intensely than the tropics, because ice melts in the Arctic and the bare ground absorbs more sunlight than the ice did. You can see it in Arrhenius' results, you can see it in the Arctic climate records of the past few decades, and you can see it in the global warming forecast for the future. Full disclosure: where you can't see it is in Antarctica. It's a bit of a mystery how cold it's been in Antarctica; it may have something to do with the ozone hole.

Another amplifying feedback to global warming involves water vapor. Water vapor is a greenhouse gas, responsible for capturing more of the outgoing infrared radiation in the atmosphere than CO_2 does. The fact that water vapor is a stronger greenhouse gas than CO_2 does not mean we needn't worry about rising CO_2 concentrations. The concentration of water vapor in the atmosphere is controlled by the fact that if the humidity gets too high, it rains. Warmer air can carry more water vapor than cool air can, so warming from rising CO_2 could lead to more water vapor in the atmosphere. Water vapor warms the Earth still further, because it is a greenhouse gas.

Like the ice albedo feedback, the water vapor feedback is an amplifier of global warming. Unlike ice albedo, which is confined to high latitudes, the water vapor feedback has a rather more uniform effect around the globe, and it about doubles the temperature change we expect from rising CO_2 alone.

There is uncertainty about how strong the water vapor feedback is. The question is whether a warmer world could be drier or wetter than we expect it to be. The average relative humidity of the Earth's surface is about 80%. Arrhenius assumed that the atmosphere would remain 80% relative humidity as it warmed. A relative humidity of 80% represents more actual molecules of water in the warm atmosphere than in the cooler atmosphere, because warm air holds more vapor than cool air. Modern climate models also predict that the relative humidity will not change much with global warming. If the real atmosphere turns out to get wetter with rising CO_2 than models predict, for example, the real water vapor feedback would be stronger than we expect.

Though the answer hasn't changed much, the quality of the answer has certainly improved in the last century. Many pieces that Arrhenius simply had to guess at can now be predicted based on a mechanistic understanding of how things work. Just as important, the models that make the predictions have been tested against reality. In the 1930s, scientists were excited by a theory that sunspots controlled climate by changing the intensity of the Sun. A prediction was made, based on the weather patterns of the recent past, that it ought to get drier in Africa during the sunspot minimum of the 1930s. It turned out that Africa got wetter during the sunspot minimum, so that was it for sunspot theory. The intensity of the Sun is currently thought to have a large impact on century-timescale climate fluctuations such as the Medieval Optimum and Little Ice Age climates, described in Chapter 3. But variations in solar intensity in the last few decades have been weak compared with the change in climate forcing from greenhouse gases.

One problem that might seem like a show-stopper for climate forecasting is the discovery in the 1960s by Edward Lorenz that the weather is fundamentally unpredictable beyond a time horizon of a week or two. One popular name for this phenomenon is "chaos," and another is the "butterfly effect." The idea is that two nearly identical states of the weather, differing only a little bit, will tend to diverge from each other, so that a small initial difference between the two will grow with time. Small imperfections in a model of the weather today will grow, until eventually all that is left in the model is amplified garbage. The weather forecast for tomorrow is pretty good, and my impression is that the forecasts have been getting better every year. The weather forecast for 10 days from now however has always been and continues to be pretty much useless. How can we expect to forecast the weather in 100 years, let alone in 100 millennia, if we can't do 10 days?

The answer is that no one is attempting to forecast the particular weather to expect on a particular day a century from now. Individual fluctuations of weather are chaotic, but the time-averaged weather, called climate, is not. Drawing once again on our sink analogy (Figure 1), waves on the surface of the water could be called weather, while the average water level in the sink would be climate. Climate is constrained by the simple, system-wide energy budget just as the water level in the sink is constrained by water through-flow. Predicting weather would be like predicting the waves in the sink, which requires that you know a lot more about the water in the sink than just the through-flow.

Climate models have their own weather, which is a useful estimate of the statistics of future climate, the frequency of storms and things like that. And long-time averages, say a temperature average over 10 Januaries, can be compared between the model and the real world. Perhaps this is as good a definition for the word "climate" as any; those aspects of the weather that can be predicted far in the future, in spite of the fact that weather is chaotic.

The energy budget of the surface of the Earth varies from place to place, because the temperature varies from place to place. The sink in our analogy only had one water level, but the Earth has a range of surface temperatures. The energy flowing in as sunlight might get transported to another location by the winds or ocean currents before it is lost to space as infrared.

Energy tends to be exported from the tropics, where the sunlight is most intense, to higher latitudes. The high latitudes act like cooling fans of the planet, keeping the tropics cool by carrying heat away and helping to ditch it into space. Dramatically, if the tropics were isolated from the high latitudes, unable to use the poles as cooling fans, the oceans in the tropics could boil in a phenomenon called a runaway greenhouse effect. We are not in danger of experiencing a runaway greenhouse effect on Earth, but it happened on Venus.

Heat is carried around the surface of the Earth by fluid flow, which is tricky to simulate or understand. Simple, slow, gooey flows like molasses are fairly easy to describe, but when the flow becomes turbulent, forget about it. Turbulent flow is one of the great challenges for computation, because there is interesting stuff going on at a huge range of spatial scales. On Earth, circulation patterns range in scale from millimeters up to the size of the Earth.

Some phenomena in nature, such as the ballistic arc of a baseball, can be described pretty well by simple equations. Unfortunately, there are no simple equations that capture most types of fluid flow. Fluid flow can be simulated on a computer by chopping the domain of the problem (the Earth's atmosphere or oceans) into pieces or blocks on a 3-dimensional grid. Each block has a single temperature, one wind velocity, a water vapor content, etc.

We can stretch the sink analogy to correspond to our multi-temperatured Earth, although the analogy is starting to get contorted. We would need an array of sinks, each with drains, and separate faucets, and water that would be allowed to flow from sink to sink. Each sink could have a slightly different water level from its neighbors. Some sinks would generally have more water than others, but there would also be a good deal of sloshing around (weather).

Climate science is climbing a brick wall in the fact that increasing the detail of the simulation makes the computer program run much much more slowly. If you want to double the model resolution, it requires twice as many grid points in each of three dimensions, equaling eight times more work to do per time step. Making matters worse, time steps have to get shorter as the grid boxes get smaller, or else the model crashes. Doubling the resolution of a calculation results in a model that will run sixteen times slower.

Clouds are probably the toughest challenge to simulate in a rigorous, mechanistic, first-principles way. The character of the cloudy skies on Earth depends on collisions between droplets on spatial scales of millimeters, on upward and downward gusting winds on spatial scales of meters, the convergence of winds on the storm scale of 100 kilometers, and on the global atmospheric circulation. Doing this right would take a lot of grid points.

The ideal thing would be to put all of this complexity into a computer model that only knows fundamentals of physics and chemistry, and have the model predict what clouds should look like. Talk about tedious calculations; this would be too much even for the fastest computer. For many years (an eternity in computer time), the fastest computer on Earth was a Japanese machine called the Earth Simulator; this machine was not nearly fast enough to resolve all of the physics of clouds and turbulence in the Earth's climate system. Even with the explosive growth in computing power described by Moore's law, computers in the foreseeable future are never going to be up to doing the calculation that climate scientists would be most happy with.

Plan B is to program into the model the large-scale behavior of, say, the cloudiness of the atmosphere, based on observations of cloudiness. Each grid cell box in an atmosphere model keeps track of the temperature and vapor content of the air in it, as well as the number of cloud droplets per cubic meter, the total water content, and maybe something about the size distribution of the droplets. The cloud subroutine makes a guess about how much water evaporates or condenses in each time step, and how the droplets coalesce. The scheme does not rely solely on the underlying fundamental mechanisms for the process, as would be ideal, but rather tries to capture the observed behavior in a more made-up way. The name for this approach is parameterization.

The law of supply and demand, in economic models, could be described as a parameterization. Supply and demand curves describe an emergent behavior of an economic system, a description of a result rather than a fundamental mechanism. The fundamental mechanism in this case would have to do with individual investors, which would have to be simulated in the computer program, gloating and gnashing their teeth and feeling envy and fear and greed and social ambition and reading their horoscopes. Climate may seem computationally intractable, but it is much easier to model climate than it is to model economics.

A climate scientist might come up with a scheme to describe clouds, and see that it captures the variability of the real Earth. The real earth does span a wide range of variation, from the tropics to the poles, deserts to jungles, mountains and plains. If the scheme is able to predict all of the variations in cloudiness on Earth today, then perhaps it will also capture any change in cloudiness as Earth's climate changes. However, because a parameterization is not built from the ground up using only the fundamental building blocks of physics and chemistry, it comes with no guarantees that it will change realistically if the climate upon which it's based changes too much.

There are many different ways to cook up a parameterized cloud, and it is done differently in different models. Some parameterizations are better than others. Often the best reassurances that these parameterizations are not deluded come from formal model intercomparison projects. In the 2007 IPCC Scientific Assessment, there are intercomparisons between nineteen different climate models, each developed by separate, competing groups of scientists. The models are also compared with measurements from the real world, present-day measurements or inferred climate parameters from the past. In practice the "duplicate, compete, and compare" approach seems to function fairly well at rooting out mistakes and bias.

Uncertainty in the science of climate change is often used as an argument not to worry about global warming. That logic makes intuitive sense if one thinks from a reference point of an unchanging climate. The forecast says "warming" but it could be wrong, therefore there might not be warming.

But it is known with certainty that CO_2 affects the climate. If Fourier's greenhouse effect were wrong, Earth's natural climate would be much colder than it is. It is certain also that CO_2 levels in the atmosphere are rising. The response to rising CO_2 is certainly some degree of warming; no effect or a cooling effect can be ruled out.

So the forecast calls for warming, but the warming could be more or less than the forecast calls for. In general, past climate changes, described in Section 2 of this book, are more intense than we would have expected. The future could also be worse than expected. Uncertainty in the climate forecast, when we think about it carefully and honestly, is no argument for complacency.

CHAPTER 2

We've Seen It
with Our Own Eyes

Thermometers have been around in something like their modern form since the first mercury thermometer of Gabriel Fahrenheit, in 1724. There are many stories about the origin of the Fahrenheit temperature scale, but one is simply that 0°F was the coldest temperature that Fahrenheit experienced in the winter of 1708–1709, and 100°F was his own body temperature. The scale puts the melting and boiling points of water at the somewhat awkward temperatures of 32°F and 212°F, respectively. We in the United States tend to think of the metric system as some newfangled thing, but the Celsius scale is no younger than the Fahrenheit scale. Fresh water melts and boils at 0°C and 100°C on the Celsius scale.

The very first temperature measurements might not have been easily comparable with each other, while the temperature scales were being defined. But the modern definition of the Fahrenheit scale was chosen just a few years after Fahrenheit's death in 1739. A mercury thermometer is still considered a good way to measure temperature today.

Temperature is a relatively easy physical quantity to calibrate, because common substances like water have precisely stable melting and boiling temperatures, the same temperatures every time even centuries apart. Other climate variables, such as the atmospheric CO_2 concentration, were much more difficult to measure reproducibly, so direct reliable measurements had to wait over two centuries before they were made. Thermometer temperature measurements can be compared quite confidently across decades or centuries, because we all have the same water to calibrate with.

The easiest, first way to look for global warming is to compile the grand average temperature of the surface of the Earth, taken over summer and winter, day and night, tropics and poles. This average is straightforward enough to obtain from a computer simulation of the climate. The model has temperatures at regularly spaced grid points around the world and through time, and the computer can just average them up.

In the real world, the distribution of locations where the temperature is known is not as convenient as the distribution in a model, but the data coverage is still pretty amazing. The reason is that surface temperature needs to be measured carefully, everywhere around the world, in order to forecast the weather. A lot of guys in suits have been willing to pay serious money, for a long time, to know whether to carry umbrellas each day. The result is an excellent set of atmospheric weather data, records through time of temperature, rainfall, winds, humidity, cloudiness, and so forth. As an oceanographer, I can only envy the excellent quality of all this atmospheric data.

To compute the average of the thermometer readings from all over the globe, the data must be corrected for sources of bias. If there are more measurements in the daytime than at night, or more in Europe than in the Himalayas, the average must be balanced to reflect this. When there are arguments about climate trends, they usually center around potential subtle biases hidden in the data rather than arguments about the calibration or political leanings of any given thermometer.

One oft-discussed issue with regard to the reconstruction of average temperature is called the urban heat island effect. Paved land is measurably warmer than vegetated land, no doubt about it, because vegetated land cools by evaporation. The question is whether any warming in the computed average temperature could actually be the urban heat island effect instead of global warming. Hot urban centers are part of the Earth, and they do contribute to the average temperature of the Earth, but their warmth is not caused by rising CO_2 concentration.

The easiest solution is to throw out urban data, by picking it out by hand, to leave the average temperature of the non-urban Earth. This is a subjective, imprecise task, but replicate studies all find that it makes little difference to the global average whether urban areas are excluded or not. It turns out to be a non-issue. Independent, competing studies produce very similar-looking global average land temperature records, regardless of how they deal with urban heat island effects (Figure 4). So unless someone comes up with believable proof that the urban heat island is important, we'll not worry about it.

Seventy percent of the Earth's surface is covered by ocean. The temperature of the surface ocean has a lot to say about the temperature of the air just above it, unless there is sea ice insulating the two from each other. If the air is warming, the surface ocean should be warming, too. Therefore, the temperature history of the surface ocean serves as an independent check on the land temperature reconstruction.

Historical trends in sea surface temperatures also have to be corrected for biases, which turn out to be a bigger deal than the potential urban heat island bias turned out to be for the land temperature trends. Sea surface temperatures were measured in the first half of the century by lowering buckets down to the

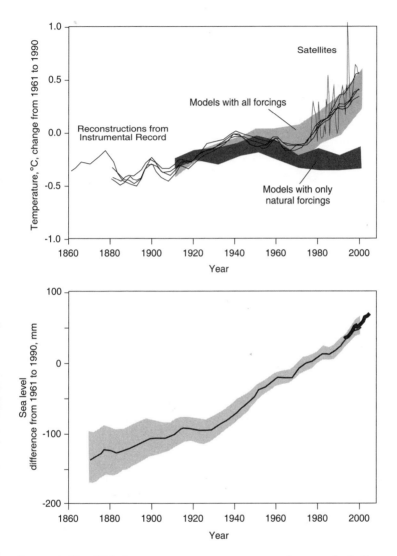

FIGURE 4. Top. Global average temperature, as measured meteorologically, by satellites, and as forecast by climate models with and without anthropogenic climate forcings. The models can capture the trend but only by admitting that CO_2 is a greenhouse gas that is changing climate. Bottom. Sea level change over the past century.

water on ropes, and then sticking a thermometer in. It's often very windy out on deck, and water on the outside of the bucket evaporates, cooling the water and biasing the temperature record a bit cold. Beginning in the middle 1940s, sea surface temperature was measured in the engine rooms of ships, as surface water is sucked in to cool the engine. These measurements, as it turns out, are closer to the real temperature of the ocean surface than bucket measurements are.

The ocean data corroborate the land data's story that the Earth is warming. The ocean surface is warming more slowly than the land surface is. This is because there is an unlimited supply of water to evaporate from the ocean, whereas the land can dry out. The same process is responsible for the urban heat island effect, as urban land dries out more quickly than vegetated land. The surface ocean is also kept cool by the huge water mass of the deep ocean, which is absorbing heat that would otherwise warm the atmosphere. The next chapter will explain that almost half of the warming we anticipate, from the current atmospheric CO_2 level, hasn't happened yet.

Satellites can measure the temperature of the lower atmosphere, by measuring the intensity of microwave light emitted by oxygen molecules. The oxygen emits microwave light more intensely the warmer it is. When the satellite looks down, it sees microwave light coming from oxygen all through the atmosphere, not just at the Earth's surface. Using various spectral tricks, the microwave signals can be converted into temperature estimates from various altitude ranges in the atmosphere.

The long-term satellite temperature record is constructed from a series of satellites that must all be calibrated the same, such that they all would agree on a surface temperature if they were all in the same place at the same time. Over time (about a decade) since the satellites first started flying, biases, bugs, and errors have been eliminated from the satellite temperature record. Some of the biggest problems had to do with changes in the readings of the instruments as the orbits of the satellites slowly wound down. At the time of the 2001 IPCC Assessment Report, the interpreted satellite record did not show the warming seen in the surface temperature records. In the 2007 report this discrepancy was resolved (Figure 4). It turns out there was a sign error someplace in the analysis.

The average temperature of the surface of the Earth has risen overall through the past century. There was an interval of cooling, from the 1940s to the 1970s, and very strong warming since then. Of the 21 hottest years on record, 20 of them have taken place in the last 25 years. That last uptick stands accused of being global warming.

Other temperature records corroborate the warming of the last decades seen in the land, surface ocean, and satellite temperature records. The subsurface ocean, for example, is a good place to look for global warming. The ocean has the capacity to store a lot more heat than the atmosphere does, and so it takes the ocean much longer to warm up or cool down. Temperature records from the deep ocean therefore emphasize long-term trends in the atmosphere, by filtering out some of year-to-year variability.

Temperatures in the subsurface ocean have been rising measurably over the past few decades. The temperature changes are largest near the surface, and they can be measured to several kilometers depth in some parts of the ocean. The deepest waters of the ocean have not warmed much at all yet.

As the ocean warms, it absorbs heat from the atmosphere, temporarily keeping the Earth surface cool. By measuring how much heat the ocean is taking up today, it is possible to estimate how much warmer the Earth will get when the ocean warms up as much as it is going to and stops taking up heat. The Earth's surface has warmed by 0.7°C since 1950, and the projection is that if atmospheric CO_2 stopped rising today, the warming would continue to about 1°C a few centuries from now.

Glaciers are melting all around the world. Most glaciers flow from some kind of valley or bowl up in the mountains where snow accumulates. The ice in a glacier begins to melt when it reaches warm air at lower elevation. When the climate warms, glaciers tend to get shorter, melting up from below. Glaciers have been melting since the end of the Little Ice Age, three centuries ago (Chapter 4), but the rate of melting has accelerated in the past decades. The snows of Kilimanjaro are projected to be gone by 2020, and Glacier National Park in the U.S. state of Montana is projected to lose its last glacier in a few decades.

Sea ice is melting, in the Arctic in particular. The decrease in the area of ice cover has been faster than any model had predicted. Summer sea ice is projected to melt completely by the year 2050. Shipping companies are happily making plans to exploit the fabled Northwest Passage, a reality at last after three centuries of searching. Polar bears without sea ice face near-certain extinction.

The Arctic Ocean covers a large area of the Earth's surface, nearby the climate-critical Greenland Ice Sheet and the deep water formation regions in the North Atlantic. Sea ice is some of the most reflective stuff on Earth, and open ocean some of the least reflective. Sunshine in the summertime Arctic is some of the most intense on Earth, if you average over 24 hours, because the Sun never sets at night. Melting of the Arctic sea ice would be a deeply fundamental change in the Earth's climate system, the impacts of which I don't believe climate models can predict very confidently. The melting of Arctic sea ice is the clearest example, to my mind, of a tipping point in global warming.

Sea level is rising (Figure 4). Two-thirds of the sea level rise today is caused by thermal expansion of the warming ocean. Melting glaciers contribute most of the rest. The major ice sheets in Greenland and Antarctica will contribute massive amounts of water to the ocean, eventually, but their contribution to present-day sea level is small (more in Chapter 12). All of the processes that contribute to sea level rise are slow, ensuring that sea level would continue to rise for several centuries even if the CO_2 concentration in the air stopped rising today.

Hurricanes appear to be getting more intense, in particular in the North Atlantic Ocean. The storm intensities, tabulated from year to year, correlate with the variations in the temperature of the sea surface, with warmer waters brewing fiercer storms. It is impossible to confidently predict the future of hurricanes, but if recent trends continue it would be very bad news.

The human impacts of global warming have been mostly subtle so far. There are exceptions, for example the Arctic has warmed intensely, and sea level rise is not subtle for some islands in the tropical Pacific. But globally, climate change has not caused the global economy to crash, or led to huge numbers of clearly identifiable climate refugees. The strongest impacts of climate change may have been in the form of extreme weather events like the European heat wave in the summer of 2003, which was said to be a once-in-five-century event, but which was repeated in 2006. The climate changes we have seen so far are much smaller than the forecast for the coming century, explained in the next chapter.

There are four external factors, agents of climate change called climate forcings, that can warm or cool the climate. (1) Greenhouse gases are the obvious one. Another is (2) sulfur from coal burning, which forms a haze in the atmosphere reflecting sunlight back to space to cool the Earth. And two natural climate forcings are (3) volcanic eruptions and (4) changes in the intensity of the Sun. Records of past changes in these climate forcings have been pieced together from measurements in ice cores.

The different climate forcings can be compared with each other in terms of watts per square meter, or watts/m^2. A volcanic eruption could decrease sunlight by one watt/m^2, or an increase in greenhouse gas concentration could decrease outgoing infrared energy by one watt/m^2. Jim Hansen, an outspoken climate scientist working for NASA, equates a watt of power with a decorative colored light bulb, and says that the greenhouse effect from our high-CO_2 atmosphere is equivalent to two such light bulbs shining down from the sky over every meter of the Earth surface. Imagine the advertising possibilities!

The largest of the four horsemen of climate change is the change in greenhouse gas concentrations in the atmosphere. Rising CO_2 accounts for just over half of our total greenhouse gas climate forcing, with the rest coming from methane, Freons, and other trace gases. Atmospheric measurements of greenhouse gas concentrations go back fifty years, and concentrations from before this time are measured in bubbles trapped in ice, mostly from Antarctica. The total change in climate forcing from human-released greenhouse gases is about 3 watts/m^2.

The other large human-caused climate forcing agent is a sulfuric haze from coal combustion. Coal contains sulfur, and when it is burned, the sulfur eventually reacts with oxygen in the air to form sulfuric acid (battery acid). The sulfuric acid forms tiny droplets in the air, called aerosols, which are very efficient at scattering visible light. The sunlight scattering effect of the haze is to cool the Earth, a partial counterbalance to the warming effect from the greenhouse gases.

Sulfur emissions also have an indirect affect on climate by changing the properties of clouds. Concentrated sulfuric acid is anxious to dissolve in water. Droplets of acid in the air tend to scavenge water vapor from the air, growing into larger, more dilute drops. Sulfate aerosols have the ability to create liquid water cloud drops where none would exist in clean air. This phenomenon is at work in contrails. Ships, too, often leave behind a line of clouds in an otherwise clear sky. The sulfur ultimately rains out as sulfuric acid, the main component of acid rain.

Aerosols also tend to decrease the droplet size in clouds. Smaller drops may survive longer in the atmosphere before raining out, making the Earth cloudier and therefore cooler. Small drops are also more efficient at scattering incoming sunlight, while large drops tend to absorb the light. This is why dark clouds are the most likely to rain; they are made of larger, heavier drops, while bright clouds are made of smaller drops. The effect of aerosols on a cloud is to make it more reflective. Dirty air (containing aerosols) cools the Earth by scattering light, and dirty clouds do more of the same.

The total change in Earth's energy budget from aerosols, in watts/m^2, is about -1 to -1.5 watts/m^2, where the negative number implies cooling. The aerosols therefore apparently cancel out a significant fraction of the warming from greenhouse gases. The energy industry has made great strides in reducing sulfur emission in most of the developed world, and as the rest of the world prospers, they will probably want to clean up their acid rain problems also. The loss of the aerosol cooling would tend to allow more warming from the greenhouse gases.

Some have proposed using the cooling effect of aerosols for deliberate cooling of the Earth. The aerosols would reside in the atmosphere longer if they were released at high altitude, in the stratosphere just above where commercial airplanes fly. The stratosphere is dry, so no raindrops would carry the sulfur back to Earth.

The longer lifetime of aerosols in the stratosphere would probably mean that the Earth could be cooled without releasing so much sulfur that our fresh waters would be poisoned with acid rain. However, the aerosol lifetime of a few years is much shorter than the CO_2 lifetime of thousands of years. Deliberately cooling the Earth using sulfates would be an ongoing project, not a one-time fix. This and other "geo-engineering" schemes will come up again in the Epilogue.

The other two climate forcing agents are volcanoes and variations in the intensity of the Sun. Volcanoes inject a sulfur haze into the stratosphere, where it deflects sunlight back to space, in the same way that aerosols from power plants do, cooling the Earth. The climate forcing from a large volcanic eruption can be as strong as about -10 watts/m^2, higher than any of the other climate forcings.

The climate impacts of a volcanic eruption are weakened considerably by the fact that aerosols in the stratosphere survive only for a few years before settling out. The most recent and best documented large climate-changing volcanic eruption was Mt. Pinatubo in the Philippines, in 1991. Four watts/m^2 of solar dimming caused 0.6°C of cooling that lasted for about two years. The event provided a natural experiment for testing climate models.

Solar variations are the smallest of all, typically on the order of 0.1 watts/m^2. The solar intensity varies on the time spans of decades and centuries: long, slow flickers in the fires of the Sun. The heat output from the Sun correlates with the number of sunspots, the visible manifestation of magnetic storms that inhibit the energy streaming from the Sun. Even though the sunspot areas themselves are cooler than average, the overall temperature of the Sun is higher when there are lots of sunspots.

People began observing and recording the numbers of sunspots in the time of Galileo in the 1600s. Just after they started paying attention, sunspots disappeared between 1645 and 1715, a period now called the Maunder minimum. This time period coincided with a period of general cool climate, at least in Europe, called the Little Ice Age (Chapter 4).

The intensity of the Sun further back in time can be estimated by measuring the products of cosmic rays, depositing in ice cores. When the Sun is brightest, it has a strong magnetic field, which shields the Earth from cosmic rays. The cosmic rays, when they reach the atmosphere, produce radioactive elements like beryllium-10 and carbon-14. A brighter Sun means less cosmic rays reaching the atmosphere, and so less carbon-14 and beryllium-10 in the ice core.

These records show minimum solar intensity during sunspot-free times like the Maunder minimum, and have informed us of solar variations even further back in time. Unfortunately, there are no direct measurements of solar intensity during times like the Maunder minimum. It requires some guesswork to translate ice core beryllium-10 and carbon-14 data into a record of the intensity of sunlight. The real variability in solar forcing could have been somewhat higher or lower than the best guess.

When the Sun gets brighter, it is the ultraviolet light that brightens the most. A watt/m^2 of ultraviolet light could have a stronger climate impact than a watt/m^2 of visible light. Ultraviolet light generates ozone in the stratosphere. Ozone is a greenhouse gas, which determines the temperature of the upper atmosphere. Change the UV, it changes the ozone, changing the air circulation and the climate. It could be that Sun is a somewhat stronger climate player than its simple watts/m^2 radiative forcing would have us believe.

Could the forecast for global warming be wrong? There are uncertainties, after all. The variability and climate impact of solar flickers are uncertain, as is the effect of aerosols on clouds and climate. Clouds are not simulated from first principles in climate models, because the computer hasn't been built that could simulate all of the gusts and droplets in the global atmosphere. And the world is a wondrously complicated and subtle place. Could there not be some phenomenon undreamed of, something un-discovered that will change everything?

Climate models are able to simulate the temperature trends from thermometers and natural proxy records if the models are subjected to all four horsemen of climate forcing, natural plus human-caused (Figure 4). Remove the human-caused forcings, and the natural forcings can't do it anymore. Solar variability, clouds, aerosols, ozone—none of these things can explain the warming of the past decade or two. The Sun has not been getting brighter. It hasn't been getting less cloudy. Ozone and UV changes haven't made the world warmer. This stuff was being measured, and it wasn't happening. The only factor driving a large warming is greenhouse gases.

But isn't it possible that some phenomenon undreamed of is responsible for the warming in the past decades? This as it turns out is a tall order. Regional climate changes could just be natural variability, the waves on the surface of the sink in our analogy from Chapter 1. But the whole world has warmed up in the last thirty years—land and ocean alike. The excess heat energy had to have come from an imbalance in Earth's energy budget.

There aren't that many ways to get energy into and out of a planet. There is visible light, which can alter climate if the albedo of the Earth is changed, for example by a change in cloudiness or ice or land cover. And there is infrared radiation, which drives climate by means of the greenhouse effect. It would not be so easy to slip energy between the Earth and space without anyone noticing or detecting it, after all these years of looking. But the natural world is a complicated and subtle place. New discoveries remain to be made, no doubt about it. For the sake of argument, suppose a phenomenon undreamed-of exists that caused the ob-served buildup of heat.

But we already have a satisfactory explanation for the warm-ing in the rising greenhouse gas concentrations. Shifting the blame to something else would require an explanation of why the CO_2 would not be trapping the heat as we expect it would be doing. Think of it like a murder mystery. The butler (CO_2) was caught with a smoking gun in his hand in the room with the dead guy. There is a lot of public interest in this case, so your boss is driving you nuts writing reports and such like; everything has to be pinned down on this one. Yes, the bullets came from the gun. Yes, the gun was purchased by the butler. Everything checks out.

But now your partner Bob argues that it was really the chauffer did it. Actually, you find out that the chauffer was at his sister's wedding on the other side of town for the whole time and lots of people saw him. But Bob says, maybe there is some way he did it but you're just not smart enough to figure it out. OK, you retort, but if Bob is going to convict the chauffer, he has to think of a way to unconvict the butler. He would have to come up with an innocent explanation for the butler's smoking gun, and the bullets and all that.

CO_2 and other greenhouse gases can easily explain the observed warming. Predictions of the effect of CO_2 on climate haven't changed much in one hundred years of climate science. For the global warming forecast to be wrong, the climate needs to be insensitive to CO_2 and other greenhouse gases. In this world we could dump as much CO_2 into the air as we like and it won't warm up very much. The new theory would need to provide a good reason to toss out the well-settled climate effect of greenhouse gases.

Note that a second bridge has just been crossed. For the global warming forecast to be wrong, two phenomena undreamed of are required, one to cause the warming, and the other to take that privilege away from greenhouse gases. This is a tall order.

The bottom line is that there are no competing theories or models for climate that can explain the climate record but do not predict serious global warming. The range of uncertainty that we have about the real world does not encompass the possibility that there will not be global warming from continued CO_2 release.

In 1990, the Intergovernmental Panel on Climate Change, IPCC, predicted that global average temperature would increase, and that global warming would be detectable above the noise of natural climate variability, by the year 2000. The call came early, in 1995, when IPCC declared a "discernable human influence on global climate." Unlike in the 1930s, when African drought was proposed to be correlated with sunspots, the prediction didn't fail this time. It just worked and has kept on working. The most recent report in 2007 concluded that it is 90–99% likely that "most of the observed increase in globally averaged temperatures since the mid-20th century is due to the observed increase in anthropogenic greenhouse gas concentrations."

CHAPTER 3

Forecast of the Century

Before we venture out into deep time, let's look in on the forecast for the next one hundred years. Our interest is not totally selfish; a lot of the action will actually take place on timescales of centuries.

The fossil fuel era could potentially last until about the year 2300, when coal begins to run out. After the CO_2 is released to the atmosphere, it takes a few hundred years, perhaps a thousand, for the CO_2 to dissolve in the ocean, as much as is going to. The atmospheric CO_2 concentration will spike upward and relax back downward, on a timescale of centuries. When this centuries-long climate storm subsides, it will leave behind a new, warmer climate state that will persist for thousands of years. That's the basic outlook.

If CO_2 emissions continue and climate responds as expected, then the surface of the Earth will be about 3–5°C (5–9°F) warmer by the year 2100. This doesn't sound all that impressive, really, on the face of it. The daily cycle of temperature is larger than this, and so is the seasonal cycle. A change in the long-term average, however, is a very different thing than a cold morning or a warm day. The climate in my home city of Chicago is expected to come to resemble that of present-day Texas or Arkansas by 2100. That sounds noticeable to me.

We will get a better idea of what a 3–5°C warming forecast means when we compare it with natural climate changes recorded in the past, by looking at ancient glacial cycles and back even further into deeper geologic time. This is the direction we will go in the second section of this book. Apropos of the third section (the future), it is worth pointing out here that the warming predicted up until 2100 is only the beginning. It takes centuries for warming to catch up with atmospheric CO_2 changes, so there will be further warming "in the hopper" even if CO_2 emissions are stopped.

Is warming necessarily a bad thing? People travel to Florida for vacation; now climate change is bringing the warmth of Florida to me here in Chicago. The effects of a small amount of warming, such as what we have already experienced, or a bit more, are subtle and some may be beneficial. Plants have a longer growing season, and they grow faster when atmospheric CO_2 is higher. But as temperatures rise further, the impacts are expected to become stronger, and more of them will be clearly harmful. The worst effects may have to do with changes in rainfall or sea level or storminess, rather than temperature itself.

Many days of the year here in Chicago could do with a bit of warming, but one clearly negative impact of a rising average temperature is an increase in the number of really oppressive hot days in summer. And of course, Chicago is a temperate city, far from the tropics. A friend of mine who grew up in India once said, "When the temperature goes above 40°C, you just don't want to eat anything." People survive, of course, but no one likes it when the temperature is too warm. The human species simply has a limited temperature tolerance.

There are limits on the cold end also, to be sure. Svante Arrhenius, the Swedish chemist who first estimated the climate sensitivity to rising CO_2 concentrations in 1896 (Chapter 1), wrote from his native Stockholm that a little warming might be pleasant. But I have another friend, a Norwegian, whose heart breaks as he watches his beloved winter skiing snow melt away.

It will be a rainier planet, overall. Among the different aspects of the global warming forecast, this one is fairly robust. The rate of rainfall is expected to increase because warm air carries more water vapor than cold air does. The global increase in rainfall is forecast to be about 3–5%. As with temperature, this change in rainfall seems small, and perhaps it is. An increase in rainfall is probably preferable to a decrease in rainfall anyway, if we had to make a choice.

However, with an increase in rain comes an increase in extreme rainfall events, strong storms that dump a lot of water in a short time. Hard rains lead to floods. Also, in spite of the general increase in rainfall, there is an increased risk of regional droughts, decade- or century-timescale shifts in rainfall patterns around the world. Some areas may just get dealt out for a while. Continental interiors are expected to dry as they warm up, potentially threatening breadbasket regions around the world, regions such as the great plains of North America. Dry desert bands of the Earth, located at about 30° N and S latitude, are also expected to dry.

The greenhouse climate has the potential to produce what are called mega-droughts, lasting for a decade or longer. Droughts of a year or two can be endured by relying on stored food and water, but when a drought lasts longer than this, reserves run out. Extended drought changes the vegetation and the soils in ways that tend to "lock in" the drought conditions.

Unfortunately it is difficult to forecast droughts reliably, so they are difficult to prepare for. Different climate models may agree about the global average climate changes, but disagree more about regional changes like droughts. Real droughts in the past are generally more extreme than are droughts in models, perhaps because of feedbacks, such as between soils and vegetation, that are missing or too weak in the models.

Melting mountain glaciers in the Himalayas supply fresh water to over a billion people near the Ganges, Indus, Brahmaputra, Salween, Mekong, Yangtze and Huang He rivers. Mountain snow holds the winter snowfall, releasing the water in spring and summer, conveniently when the agricultural need is greatest. The glaciers that provide summer water are also melting in the Peruvian Andes, and in the Pacific range mounts in the American northwest. The loss of the glaciers may put a serious dent in the water supplies to these areas.

Warming of sea surface temperatures may drive an increase in the intensity of tropical cyclones, also known as typhoons or hurricanes. A tropical storm requires a trigger in order to get started—a convergence of winds in a frontal system, perhaps. Every year there are 80–90 tropical storms that form around the world, only a fraction of which graduate into full-blown hurricanes. Once the tropical storm is started, it either grows into a hurricane, or doesn't, depending primarily on the temperature of the ocean water, and on whether the winds in the atmosphere will leave it undisturbed, or tear it apart.

Tropical storms tend to intensify as the sea surface warms, everything else being equal. Hurricane intensities reconstructed from satellite images seem to show strengthening as sea surface temperature warms over the past decades. In fact the storms are getting stronger even more rapidly than current theory predicts.

The National Ocean and Atmosphere Administration officially attributes the rise in hurricanes to a natural cycle called the Atlantic Multidecadal Oscillation, which is an oscillation in Atlantic surface ocean temperatures. However, there is a clear global warming imprint on sea surface temperatures, which are warmer now than they were in the last positive phase in the Atlantic temperature cycle in 1940–1960. If the warming is not a natural cycle, and the storms are following the warming, then the storms must not be strictly natural either.

The forecast for the future is murky; there is a clear danger that hurricanes could intensify with warming, but it is impossible to say exactly by how much. Scientists don't understand the working of hurricanes in the climate system well enough to be able to predict how strong they will get.

Sea level is projected to rise by about 0.2–0.6 meter in the coming century. I will go out on a limb here and predict that the impacts of this sea level rise will be most noticeable in low-lying coastal regions. Miami, New Orleans, the Netherlands, Bangladesh, Shanghai, and New York stand threatened if sea level rises too much. Sea level rise has already led to plans for evacuation of natives of a few tropical Pacific islands such as Tuvalu and Vanuatu.

Two-thirds of the sea level rise by 2100 in the IPCC forecast comes from thermal expansion of sea water. It will take centuries for the temperature of the ocean to stop warming, and to reach equilibrium with a new climate. Sea level rise from expansion of seawater will therefore take centuries to play out. The other component of sea level rise is the melting of ice on land. Most of this comes from melting of mountain glaciers, smallish glaciers, and ice caps in places like Iceland and the Alps.

The large ice sheets, in Greenland and Antarctica, are projected (modeled) to contribute relatively little to sea level rise in the coming century. Ice sheet models generally agree that, given about 3°C or more of warming and enough time, the Greenland ice sheet would melt. Greenland ice holds enough water to raise sea level by 7 meters, enough to change coastlines around the world. Models of ice melting generally predict that it will take many centuries, even millennia, for Greenland to melt. They do not predict much melting in just a century.

However, there are reasons to fear that the ice models used to generate the forecast may be too sluggish to predict the behavior of real ice. There are melting events documented in climate records of the past which the models can't predict or explain either. One example is the Heinrich events of 50 millennia ago, the century-timescale crumbling of the Laurentide ice sheet into icebergs in the North Atlantic. Another is a time interval 14 millennia ago, called Meltwater Pulse 1A, in which three Greenland's worth of ice flowed into the ocean in one or a few centuries. We'll come to this topic again in Chapter 4.

If Greenland were to collapse into the ocean today the way the Laurentide ice sheet did to make the Heinrich events, there would be nothing that could be done to stop it. The resulting sea level rise might provoke the West Antarctic Ice Sheet to float from its submarine moorings. And it might change the circulation of the water in the North Atlantic, potentially changing the climate in northern Europe and elsewhere in the high Northern latitudes.

An increase in temperature in Antarctica is not expected to increase melting much, because the temperature is so far below freezing. The ice doesn't really begin to melt until it has been dumped into the ocean. When the air warms up it snows more, so the forecast for the next century is for the Antarctic ice sheet to grow. This is verified by measurements of ice thickness in Antarctica.

However, flow from the West Antarctic Ice Sheet into the ocean is funneled through a series of ice streams, which models are not very good at predicting either. An ice stream flows at a breakneck pace of several kilometers per year, through an ocean of more sedentary ice drifting downhill at a slower pace of just a few meters per year. The ice stream may be faster than the rest of the ice around it because the friction of its motion generates heat, providing melt water to lubricate the bed, stimulating even more flow. Ice streams have the potential to respond very sensi-

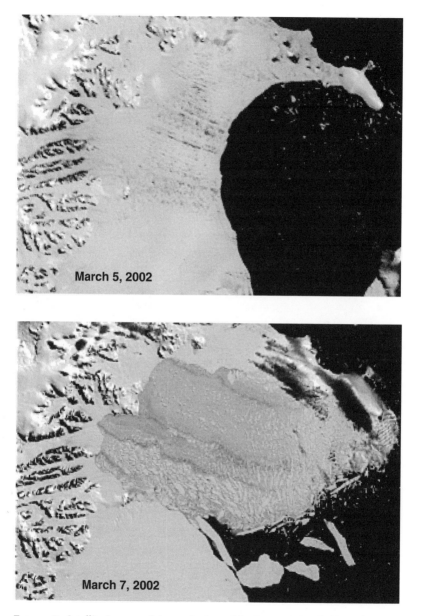

March 5, 2002

March 7, 2002

FIGURE 5. Satellite images of the explosion of the Larsen B ice shelf on the Antarctic Peninsula.

tively to changes in climate, especially if ice is already melting at the surface.

The ice streams draining the West Antarctic Ice Sheet flow into a thick plain of floating ice hundreds of meters thick called the Ross Ice Shelf. Ice shelves have also shown us a catastrophic stunt that the models hadn't foreseen. The Larsen B ice shelf on the Antarctic Peninsula exploded in 2002, converting a continuous region of ice the size of New Hampshire into a blue slurpy mash of tiny icebergs in just a few days (Figure 5). The explosion was provoked by the presence of meltwater ponds at the ice surface. The pools of standing water apparently created chasms in the ice, undermining its structural integrity. One theory for the sudden explosion of the ice shelf is that the chasms were close enough together that the pieces of ice tipped over like a long, floating train of dominos.

The demise of an ice shelf has no effect on sea level, because the ice was floating in the sea already. However, ice streams, flowing into ice shelves, begin to flow faster when an ice shelf disappears. This has been observed upstream of the melting Jacobshavn Isbrae ice shelf in Greenland and the former Larsen B on the Antarctic Peninsula. The West Antarctic Ice Sheet flows to the ocean through the Ross Ice Shelf, which is beginning to produce the same sorts of melt ponds as were seen on the Larsen B. If the Ross Ice Shelf were to explode, it might allow the melting of the West Antarctic Ice Sheet to accelerate.

In general, observations from the present day and from the geologic past suggest that ice has capability to melt more suddenly than our current models seem to give it credit for. The way to melt an ice sheet quickly is to turn it into icebergs in the ocean, and float them down to low latitudes where the sunshine is. It has happened before. More about this in Chapter 10.

Sea level rise encroaches on human welfare sometimes very slowly and other times very quickly. Agricultural land is poisoned by salt when sea level rises. The Pacific island of Tuvalu is an example of this, vividly described in Mark Lynas' book *High Tide*. Salt water rises up through the water table, slowly choking off the potential for growing anything. Because the land itself is porous, sea walls and dykes were never a viable option for Tuvalu. The islanders import food from across the ocean, and are making plans to evacuate in the coming decade. Ten thousand islanders are a movable population, a sad but manageable tale, but if the same thing happens to millions of subsistence farmers in Bangladesh or China, it is another story.

Rising sea level also exacerbates floods and storm surges. Floods can be caused by a variety of factors, some of them climate-related, such as rainfall or snowmelt, and others having to do with land use decisions (where people build houses) and river management (dams and levies). A storm surge is a lifting of the sea surface by the low atmospheric pressure inside a large storm such as a hurricane. The surge resembles a high tide, in that it comes and goes in a few hours or days. If hurricanes intensify, the higher storm surges would exacerbate the global sea level rise.

One probably robust prediction is that negative impacts of climate change will be felt most severely in under-developed countries. Casualties from a given natural assault, a storm, tsunami, or earthquake, tend to be much higher in poor countries than in rich ones. A century of global warming in the United States will probably involve uncomfortable summers, maybe some drought, and colorful headlines about hurricanes. Holland is an example of a prosperous country that has managed to build and maintain dikes, to keep the sea out of their low-lying landscape. Sea level rise in Bangladesh is less likely to be defended against, because of the length of the coastline, and the lack of economic resources to throw at the problem.

Some aspects of the greenhouse climate will be beneficial. CO_2 from the air is a nutrient for plants, and so higher CO_2 concentrations, coupled with a longer growing season and more rainfall, might be beneficial to agriculture. In general, the impacts of rising CO_2 are predicted to be mixed, positive and negative, for mild climate changes such as we have endured so far. Stronger climate changes, such as those that are forecast for the year 2100, are generally expected to be more harmful than good.

The century timescale is where practical concern gets off the bus. As far as rational self-interest goes, this is our stop. Most books about global warming end right here. But the Earth is old, and has seen many climate changes. Climate variability in the deeper past provides a context for evaluating the forecast for the future. Is global warming a big deal, or is it just nature-as-usual?

The first century is an impressive beginning, but the climate effects of global warming will persist for hundreds of thousands of years. Of course, forecasting the deep future is a tricky business. It is probably impossible to confidently predict the long-term future of human society, for example. But the release of CO_2 into the atmosphere will have a predictable long-term impact on the carbon cycle and the future evolution of climate, and we know this because of what we've seen happen in the past.

We will explore the geologic history of climate change in Section 2, setting the stage for looking into the deep future in Section 3.

Section 2

THE PAST

CHAPTER 4
Millennial Climate Cycles

One way to get a handle on the scale of the global warming forecast is to compare it with natural climate changes in the past. Climate varies naturally, for several different reasons that tend to act on different timescales. In this chapter, we will consider natural climate changes of about a thousand years' duration, and in the two chapters after that, climate changes on successively longer timescales, ultimately reaching millions of years in Chapter 6. Chapter 7 puts everything back together again, to compare climate changes in the past to those that are forecast for the future.

The first meteorological observation network was established in 1653 in Northern Italy. By the middle 1800s, weather observations including temperatures were recorded throughout the inhabited world. The earliest reliable record of global average temperature dates to about 1860.

Climate changes before this time are pieced together using what are called proxies of past climates. The first proxy method used to shed light on past climates was derived from grains of pollen preserved in lake sediments. A pine forest prefers a cooler climate than would suit a grove of maple trees, for example, so pine pollen instead of maple indicates cooler conditions. Fossil pollen can also record times of drought.

Pollen records revealed a sudden warming event at the end of the last glacial time, about eleven thousand years ago, followed by a thousand-year long cold snap called the Younger Dryas interval. An encore of the fierce glacial climate, the Younger Dryas ended as swiftly as it began, with most of the temperature change coming in just a few years. The Younger Dryas event was named for the presence of pollen from the cool-weather mountain flowers called Dryas octopetala or white dryas. Similar pollen was found in the records of the last glacial climate itself, originally referred to as the Oldest Dryas interval.

Another early technique for estimating past changes in temperature is based on analyzing the oxygen atoms in $CaCO_3$ from sediments of the ocean. Elements such as oxygen come in different sizes, called isotopes. They all behave chemically as oxygen, but they have different masses because they contain different numbers of the heavy neutral particles called neutrons. By measuring the ratio of different isotopes of oxygen, climate scientists can piece together information about past temperatures and other climate variables.

Ice sheets and glaciers also preserve traces of past climates. Core samples drilled through the ice contain actual samples of ancient atmosphere, trapped these hundreds of thousands of years in bubbles or dissolved in the ice. These samples reveal the past concentrations of CO_2, methane, and other greenhouse gases. What I as an oceanographer would give for comparable ancient seawater samples!

The abundances of isotopes of oxygen and hydrogen tell of past temperatures on the ice sheet. The temperature and atmospheric CO_2 concentrations are strikingly correlated with each other through the glacial cycles (Chapter 5), as can be observed by looking at ice cores in Antarctica. The amount of dust in the ice core tells us something about changes in rainfall and wind speed. The longest ice core records come from Greenland and

Antarctica, but there is some old ice even in the tropics, up in mountain glaciers.

Tree rings provide a highly time-detailed record of past temperatures. Trees grow better, constructing a thicker annual layer, when the temperature is higher. Tree ring width is also affected by water supply, soil conditions, grazing by deer, infestation by parasitic insects, and shading by other trees. These individual idiosyncrasies have to be filtered out of the raw data, which is done using statistical tools. The tools are "trained" to extract the historical temperature record from the tree ring data that exist for that time period. Then the older data are treated in the same way, to estimate temperatures from prehistoric times.

A puzzling fact is that tree rings don't show the warming of the past few decades. Perhaps the trees are being fertilized by the unprecedented high CO_2 concentration in the atmosphere, or the nitrogen component of acid rain. However, tree rings do correlate quite well with temperatures in the instrumental record before 1970, and agree with other climate proxies through the last millennium.

Mountain glaciers leave behind a climate record of sorts. The glacier flows downhill until it melts, further downhill in a colder climate. A glacier leaves behind a scar on the landscape called a moraine. These are piles of rocks and boulders that were carried by the ice and dropped when it melted. Most of the mountain glaciers in the world have been melting for several centuries, in response to the end of a climate interval known as the Little Ice Age. But the melting has accelerated in the past few decades in response to the recent warming.

Temperatures of the past can be reconstructed by measuring temperatures down in drill-holes or bore-holes, in ice or in rock. A change in the temperature at the surface at some point in the past can be diagnosed from the temperature in the bore-hole as it changes with depth. These so-called borehole temperatures don't contain much information about fast climate events like

El Niño, but they do remember the slower, long-term changes like the Little Ice Age.

Let's face it, none of these methods of reconstructing the climate of the pre-historic past are as good as if there had been real thermometers there. Taken collectively, however, as independent estimates, their agreement makes the end result fairly strong. Thinking about all these different methods, I have to marvel at human ingenuity. I also feel that no directed search could have been as innovative as has the entrepreneurial, competitive culture of the scientific enterprise.

Temperatures in Europe, and probably globally, were cooler a few centuries ago than they are today. The Little Ice Age lasted from about 1300 to 1800 A.D. In Europe where the cooling is particularly strong, it was at most 1°C (on average) colder than our own "natural" climate, defined as that of fifty years ago. The Little Ice Age climate was moodier, nastier, and much more variable than the climate before or since. Decades of cold were followed by decades of drought, or decades of warmth, or decades of extreme rain.

Solar astronomers in Europe and China reported a paucity of sunspots during this time. There were even a few decades, called the Maunder minimum, where no sunspots could be seen at all. In recent decades, the intensity of the Sun has varied, correlated with the number of sunspots (Chapter 2). There are no good measurements of the solar intensity during a time of no sunspots like the Maunder minimum, so it is a bit of a guess how bright the Sun was then. As explained in Chapter 3, radioactive elements like beryllium-10 and carbon-14, produced by cosmic rays, also leave a tale of past solar brightness. Both proxies suggest that the cool climate of the Little Ice Age can be explained as the result of a cooler Sun.

Before this time was a period of general warmth called the Medieval Optimum, lasting from 800 to 1300 A.D. Temperatures, at least in Europe, were warmer than in the Little Ice Age, perhaps as warm as today, more or less. This was a time of stable climate and bountiful harvests in feudal Europe, the age of gothic cathedral construction, built in praise and gratitude to a benevolent God. The Vikings were a product of the Medieval Optimum; they sailed through those same centuries, from 800 to 1300 A.D. The end of Norse occupation of Greenland coincided ominously with the onset of the Little Ice Age.

The Medieval Optimum was a time of persistent drought for much of North America. Trees grew to be two hundred years old where Mono Lake in California is now, indicating that the landscape was dry enough that this lakebed never flooded. These dry centuries were far more severe than the dust bowl drought of the 1930s. The collapse of the Classic Maya civilization, and the abandonment of the pueblos of the Anasazi, coincided with multi-year droughts as recorded in ocean sediments and tree rings.

There has been considerable wrangling in the press and even in the U.S. Congress about this question: Is it warmer now than it has been in the last 1,000 years, or was it as warm during the Medieval time? This is an interesting question scientifically, but it is an irrelevant question to evaluating global warming. If the year 1000 A.D. was as warm as the year 2000 A.D., then one might argue that our warming is natural, not an indication of global warming at all. But the Medieval Optimum warmth was probably the result of a warmer Sun, according to the solar proxies. In our time, the Sun has not been getting warmer since 1970. The warming in the past few decades can be explained only as a result of rising greenhouse gas concentrations.

One might also argue that the medieval climate was benevolent, and that therefore warming is a good thing. The world today is about 0.7°C warmer than our defined "natural" climate fifty years ago, and perhaps the Medieval climate was 0.5°C warmer than fifty years ago. Neither ought to be confused with the 3°C warmer forecast for the year 2100, which is five or six times warmer than either. Beware the bait and switch! The Medieval climate was generally benign, as is the present-day climate. But the business-as-usual forecast for the year 2100 is a different beast entirely.

There is a hint of a 1500-year cycle in climate, most strongly from the time of the glacial climate discussed in the next chapter, but also in a few climate records from the current interglacial time. What if the Medieval Optimum and Little Ice Age were pieces of such a cycle?

Climate fluctuations during glacial time, called the Dansgaard–Oeschger events described below, were strongly periodic, recurring on a precise rhythm of about 1470 years. From our current interglacial time, a few climate records show something that looks like a 1500-year cycle, but these are all from a particular region in the North Atlantic. You don't find it anywhere else. The Medieval Optimum was a real climate shift in Europe, but it was not clearly global. Other places marched to their own drummers.

But if the Medieval Warm and Little Ice Age climates were part of some reliable cycle that scientists have yet to figure out, then looking forward into the coming centuries we would predict a natural rise in temperature, as the Little Ice Age subsides into a next warm phase of the cycle. The existence of a 1500-year cycle would not make the forecast of global warming wrong. As explained in Chapter 2, the warming of the last few decades is still nicely explained by rising CO_2, and future climate is still expected to warm even more because of continued increases in CO_2. The 1500-year cycle would make the forecast worse.

The largest climate change in the last 10 millennia took place 8.2 millennia ago, in an event called the 8.2k event (Figure 6). The general climate of the Earth was somewhat warmer than today,

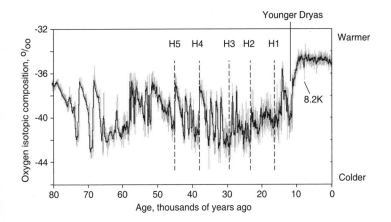

FIGURE 6. Temperature in Greenland recorded in the GISP II ice core. The dashed lines labeled with "H" are Heinrich events.

because of the configuration of the Earth's orbit around the Sun at that time (see Chapter 5).

Pollen records tell of a sudden cooling and a global tendency for drought that lasted for several centuries. All oceans carry heat from the low latitudes to high latitudes, in the wind-driven circulation of surface waters. The Atlantic Ocean is special because of an overturning circulation, in which surface water carries heat to the North, cools, and then sinks to depth. The other oceans do not carry heat across the equator, but the Atlantic does, because of this deep, overturning circulation.

The 8.2k climate event was the result of a disruption of the circulation in the Atlantic Ocean. The melting Laurentide ice sheet filled an enormous lake called Lake Agassiz, larger than all the present-day Great Lakes combined, one side of which was dammed by the ice sheet itself. When the dam broached, the water quickly made its way to the North Atlantic. Fresh water floats on top of salty water, creating a barrier to the overturning circulation like a stick caught in the spokes. This theme of fresh water killing the ocean circulation in the North Atlantic will come up again as part of the Heinrich events and the Younger Dryas, described below.

The past 10 millennia were a time when climatically not very much exciting happened. This period is known as the Holocene, from the Greek "holo," whole, and "cene," recent. The Little Ice Age, Medieval Optimum, and 8.2k events were all in the Holocene period, but they were small compared with earlier climate changes. Compared to glacial time, the Holocene has been a sunny picnic in the park. Civilization and agriculture developed during this oasis of climatic stability. Brian Fagan described the effects of the arrival of the Holocene on human civilization in a book entitled *The Long Summer.*

The Holocene is considered by some paleoclimatologists to have ended already, replaced by a new geological period they call the Anthropocene, "anthropo" the Greek for human. Geologic time periods in the past are generally delineated by major changes in climate or by biological extinctions. Earth's alleged graduation from the Holocene to the Anthropocene is therefore a statement that humankind has become a powerful force in Earth evolution.

Looking back even further in time, we begin to bump our heads against the glacial climate cycles, the topic of the next chapter. The last glacial time peaked about 21 thousand years ago, covering large areas in North America and Europe with ice sheets like those of Greenland and Antarctica today. The Earth today is in an interglacial climate stage.

Before the 8.2k event was another called the Younger Dryas, which can really be understood only within the context of the end of the last glacial time (Figure 6). The glacial cycles are driven by wobbles in the Earth's orbit around the Sun, in rhythms of tens of thousands of years, the timescales of the next chapter. However, some of the climate transitions associated with the glacial cycles however were much faster than the orbital forcing, taking place in a millennium or less, placing them in this chapter rather than the next.

Twenty one thousand years ago, Earth was in the deepest, coldest glacial climate, the Last Glacial Maximum. Sea level was 120 meters lower than today, and the average temperature of the Earth was about 5 to 6°C colder. Warming began 18 thousand years ago in the Southern hemisphere, as recorded in Antarctic ice. The temperature in Greenland remained cold, until it suddenly shot up fourteen thousand years ago. More than half of the glacial-to-interglacial temperature change in Greenland happened in just a few years. This climate transition correlates with a change in the Atlantic overturning circulation. A warm, roughly interglacial climate followed and persisted for a thousand years.

At this point another dose of fresh water from the melting Laurentide ice sheet dumped into the North Atlantic, triggering the Younger Dryas. Similarly to the 8.2k event, the result was stagnation of the Atlantic overturning circulation leading to cooling. The Younger Dryas absolutely dwarfs the 8.2k event in severity. The Earth plunged back into what was essentially a glacial climate redux, which persisted for another thousand years. The impacts reached around the world worldwide, but were strongest in the high northern latitudes.

Brian Fagan argues that the climate pulses of the deglaciation provoked the development of agriculture and collectivized human culture. In the glacial world, humans were organized in small, mobile groups that were able to adapt opportunistically to changes in the carrying capacity of the landscape. During the initial, thousand-year warm period, human society grew attached to the land, sustained by forage crops such as acorns that can be stored. When the millennial warm period ended and the Younger Dryas began, humankind was too tied to the land to move. So they learned to farm.

When we look further back in time, earlier in the last glacial climate period, the climate of the Earth was much tippier, on timescales of millennia and shorter. The clearest pictures of the action come from cores from the Greenland ice sheet. The temperature in Greenland underwent large, fast, cyclic oscillations called Dansgaard–Oeschger (D–O) events. There were 25 of these D–O events during the last glacial time. They are typically larger than the 8.2k event, but are smaller than the Younger Dryas.

The basic cycle took place on a repetitive timescale of about 1500 years. In fact, according to the dating in one of the Greenland cores, they come on a beat of 1470 years plus or minus just a few percent. No one has any idea where such clockwork regularity could come from. A D–O cycle consists of a gradual cooling terminated by an abrupt warming.

The D–O cycles interact with periodic meltdowns of the Laurentide Ice Sheet called Heinrich events. Heinrich events (Figure 6) are defined by layers of rocks and pebbles found in the sediments of the North Atlantic. The pebbles originated from the region where the Laurentide ice sheet flowed out into the ocean, in Hudson Strait. The only way to get them to the middle of the Atlantic Ocean is to float them in ice. A Heinrich event runs for a few centuries before it abates.

D–O cycles come grouped into bundles called Bond cycles, in which the severity of the cold increases from one cycle to the next. The end of a Bond cycle is marked by a Heinrich event. Heinrich events are found in the middle of the coldest D–O within the Bond cycle, and are followed by the warmest of the warm intervals. The overall Bond cycle takes 8–10 millennia, from Heinrich event to Heinrich event.

A Heinrich event must represent a catastrophic collapse of the Laurentide ice sheet into the ocean. Sea level rose by 5 meters in one or a few centuries. The mechanism for Heinrich events is still unexplained, so it is impossible to predict whether the Greenland ice sheet could do a Heinrich maneuver in the future (Chapter 11).

The causes of these abrupt climate changes have been the focus of life's work for many scientists. There were a few years in the late 1990s when practically every climate paper in the high-profile journals had the word "abrupt" in its title.

The mechanisms for these climate flip-flops are still not settled, but are thought to contain ingredients of (1) the overturning circulation in the North Atlantic and (2) sea ice. Fresh water added suddenly to the North Atlantic has the potential to stop or divert the overturning circulation cell there. Disrupting the ocean circulation changed the climate, as seen in the Heinrich events, the temperature changes at the end of the ice age including the Younger Dryas, and the 8.2k event.

Sea ice tends to amplify climate shifts by reflecting visible light (the ice albedo effect), and by insulating the air from the ocean. Air over ice can get much colder than air over open water. Sea ice was more extensive in the glacial world, explaining perhaps why the glacial climate flip-flops tend to be more severe than the only interglacial abrupt change, the 8.2k event.

Climate records on timescales of millennia and shorter look like sudden flips from one climate state to another, rather than gradual transitions across a continuous range of climate states. Most of the natural climate forcings—the variations in Earth's orbit and atmospheric CO_2 concentration—change slowly. Many of the resulting climate transitions were fast, such as the transition from glacial to interglacial temperatures in the Northern Hemisphere in just a few years. The abrupt climate cycles of the glacial climate, consisting of D–O oscillations dancing with Heinrich events, seemed to come without any identifiable climate forcing at all.

If the Earth were a simpler planet, with air and ground only, no ice, no water, the long-term future of global warming would be much easier to predict. The atmosphere has a very short memory, because it doesn't take long for the air and the ground to reach the temperatures at which the energy fluxes described in Chapter 1 balance. In the real world, the oceans and the ice sheets hold so much heat that they take thousands of years to respond fully to changing climate. As a result, they give the climate system a memory, slowing its full response to changes in climate forcing such as a change in CO_2 concentration.

However, the oceans and the ice play a more dramatic role in the climate system than as a mere set of brakes to slow down any transitions. The message from the past is that the ocean and the ice are more than passive players; sometimes they take leading roles in determining and changing the climate of the Earth.

The ocean transports heat across the equator and up to the high northern latitudes, and this circulation pattern can change within a few years. Melting sea ice can alter the reflection of sunlight to space, leading to and amplifying fast climate transitions. Ice sheets respond more slowly than sea ice, but the Heinrich events show that even they are able to change the climate on timescales of centuries.

The forecast for climate change in the coming century stands in rather embarrassing contrast to the climate records of the past. The forecast is for a gradual climate response to the change in climate forcing due to rising CO_2 concentration—a smooth rise in temperature from 0.5°C excess warmth today to about 3.0°C excess warmth in the year 2100. The IPCC forecast therefore represents a sort of best-case scenario, in that it contains no unfortunate surprises.

CHAPTER 5
Glacial Climate Cycles

The discovery of the glacial cycles was a slick piece of detective work. Earth scientists two centuries ago were not professionals in the modern sense. They were not motivated by publish-or-perish rules at universities, or competition for research grant money. Many were gentlemen of privilege, or quirky devoted amateurs, with the determination to indulge a passion of choice, in this case an interest in the history of the Earth.

Here's the case. The mountains in Switzerland are scattered with rocks called exotics, large rocks that once broke off from bedrock hundreds of kilometers away. The question is, how did the natural world move so many big rocks such long distances?

The default hypothesis was of course the great flood described in the Bible. But when you get down to actually imagining these rocks being carried in a great flood, it's not so easy to do, especially for the larger ones. When it does happen that rocks are carried along in some torrential water flow, the rocks get all rounded from tumbling, and they wind up in beds well sorted by size. The Swiss rocks in question are angular and rough, definitely not rounded.

The rocks in the Swiss mountains have another distinctive feature. Their flat surfaces are scored with long, straight, parallel scratches, or smoothly polished parallel grooves. Mountain glaciers can make these kinds of signature marks, the rocks and ice grinding past each other as the ice slowly flows.

Eventually geologists on both sides of the Atlantic agreed that their respective continents had each been repeatedly covered by sheets of ice similar to what we find today in Greenland and Antarctica. Interior Greenland and Antarctica are pretty forbidding places, and are among some of the few uninhabitable places on the land surface of the Earth. Geologists in the 1800s looked at the rocks and saw that something similar had happened to their beloved landscape, over and over. It was not so clear when the world could freeze again. The moment of realization must have been rather foreboding.

The moraines from the Laurentide ice sheet are large features in the landscape. Looking at a map of North America, one can easily imagine the Great Lakes as lobes of great, ancient ice sheets. The landscape around the lakes is lined with ridges, made of rubble piles that fringed the edges of the ice sheets where the ice melted. A bike ride in the moraines in southern Indiana is a very different experience from biking on the scraped-off plain of Chicago.

Moraines are great for tracing boundaries and making maps of the glacial world. But they leave a sketchy record of the comings and goings of ice sheets through time. They document the farthest extent of an ice sheet, but there is no corresponding trace left behind by the smaller ice sheets that preceded it. The most recent, largest ice sheet wipes out the moraines left behind by the others.

Moraines are not so easy to determine the age of, either. Before the development of radiocarbon dating in the 1950s, there was really no way to tell how old a moraine was. Since then, bits of twig or moss can be dated by measuring carbon-14, but there is always ambiguity about whether the moss or twig grew at the time when the moraine was formed, or whether it could have been dead for a long time before it got buried by the rocks.

Much better information about ice sheets waxing and waning through time can be decoded from ancient samples of ice in ice sheets, and $CaCO_3$ in ocean sediments. Both ice and $CaCO_3$ carry information in the relative abundances of the oxygen isotopes, as we discussed in Chapter 4.

Oxygen isotopes in the ocean can tell us about the amount of water frozen in ice, because heavy and light oxygen isotopes get separated somewhat by the process of growing an ice sheet. When water evaporates from the ocean, the lighter water molecules evaporate a bit faster than the heavier water molecules. Water vapor in the atmosphere is therefore isotopically lighter than the liquid water it evaporated from. When vapor condenses to make rain or snow, the heavier molecules tend to go first, leaving the vapor in the atmosphere to become ever lighter, the more vapor is condensed. The atmosphere acts like a giant still, distilling out the lighter oxygen isotopes from the ocean and then depositing them on ice sheets. When the climate of the high latitudes is particularly cold, the water vapor in the air is distilled most intensely.

Ice sheets grow at the expense of water in the ocean. The ice sheets are isotopically light, and the ocean left behind is isotopically heavy. You would never notice the heavier water if you were swimming in the glacial ocean, the way one can feel the difference in buoyancy between swimming in fresh water versus salt water. However the difference is easily measurable in the laboratory.

The oxygen isotopic composition of the ocean is recorded in samples of $CaCO_3$ that form in the ocean. Even in tropical regions where ice sheets would be the furthest thing from one's mind as one swam along, the $CaCO_3$ in sea shells growing today carry information about how large the ice sheets are at the other end of the world. $CaCO_3$ shells settle to the sea floor, leaving behind a continuous record, new ones on top of old ones, as convenient as a tape recorder. From the sediments we get a much more detailed history of the comings and goings of the great ice sheets than can ever be pieced together from clues on land such as the moraines.

The evolution of ice sheets—their growth and decay through time—seems to be paced by variations in the Earth's orbit around the Sun. The impact of the orbital wobbles is to redistribute the solar heat from place to place and season to season. The total amount of solar energy the Earth gets over the course of the year does not vary too much, but the distribution of the energy varies drastically. The changes in solar energy are reversed across the equator, so if the Northern hemisphere summer is getting a lot of heat, the Southern hemisphere summer is cool.

The Northern hemisphere is where the biggest changes in ice volume take place, the Laurentide and Fennoscandian ice sheets in North America and Europe. These ice sheets form in cold times and then melt away completely when it warms like today. The ice in Antarctica runs all the way to the ocean, today and during glacial times, and so its size doesn't change as much over a glacial cycle. The Greenland ice sheet also covers most of the available land, today and in the last glacial time, so its variations through time are also muted, although there is evidence that the ice sheet was smaller during the last interglacial climate interval 120 millennia ago. The ice sheet in North America was bigger than the present-day Antarctic, and it melted away completely during warm times like the present.

The northern hemisphere summer, of the places and times on Earth, is kind of a sweet spot for driving its climate. Changes in sunlight intensity there hit the Earth like a sucker punch, plunging the entire planet into the deep freeze of a glacial climate. It is summertime that is most important because it always gets cold enough to snow in Newfoundland in the wintertime. The question that determines the fate of an ice sheet is whether the winter snow will survive the summer without melting. And because the ice sheet in question is in the Northern hemisphere, sunlight intensity in the northern hemisphere summer in particular is the quantity to watch.

The orbit of the Earth is like a bell that rings with various frequencies, a three-part harmony. The low part is like two musical notes, a 100-millennium cycle superimposed on another cycle of about 400 millennia. This part arises from changes in the eccentricity of the Earth's orbit. Sometimes it is more circular than other times. Today, the orbit is approaching its most circular; the last time this happened was about 400 millennia ago (Figure 7).

The high note is a phenomenon called the precession of the equinoxes. The axis of rotation of the Earth, a line connecting the North and South poles, is tilted relative to the plane of the Earth's orbit around the Sun. The direction of the tilt is not always the same, but it spins around like a top winding down, taking about 20 millennia to go full circle.

The effect of precession on climate is that sometimes the Northern hemisphere summer—the place in the orbit where the North Pole leans toward the Sun—occurs when the Earth is close to the Sun, on the near part of the ellipse of the orbit. At the other side of the cycle, 10 millennia later, the Northern hemisphere summer happens in the far part of the orbit. This is where the Earth is today.

The middle note is caused by the angle of the Earth's tilt, called its obliquity. Sometimes the poles are tilted a bit more than at other times relative to the plane of the Earth's orbit. The differences are small; between angles of 22° and 25°. When the axis is more tilted, the seasons are more intense. The obliquity of the orbit varies with a period of 40 millennia. The Earth's poles today are near the minimum tilt angle, about 22.5°.

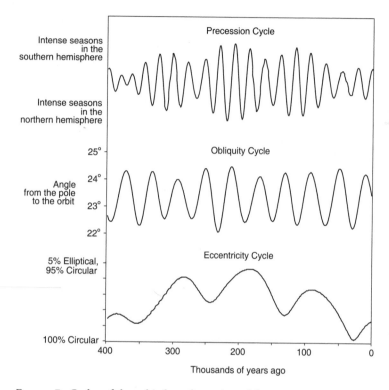

FIGURE 7. Cycles of the orbital configuration of the Earth.

In the early part of our current ice age, beginning about 2 million years ago, there was a fairly clear similarity between orbital forcing, defined as the solar intensity in the Northern hemisphere summer, and the amount of ice on Earth. The climate system was a high-fidelity record player, playing the notes from orbital eccentricity, tilt, and precession of the equinoxes, all at the relative volumes dictated by the orbital forcing.

Beginning about 800 millennia ago, the record player started to distort the orbital music. It was as if someone had pumped up the bass on the amplifier. The comings and goings of the ice became dominated by 100-millennium oscillations. Perhaps these 100-millennium cycles are caused, or paced, by the 100-

millennium cycles in the eccentricity of the Earth's orbit, but if so they are blown completely out of proportion to the strength of the 100-millennium forcing. Also there is no trace of the 400-millennium cycle note that eccentricity also plays.

The reason for the growth of the 100-millennium cycle is not yet settled, but one theory that will come up again in Chapter 12 is that quirks in the life cycle of ice sheets may be important. Once an ice sheet forms and starts to grow, it tends to keep growing until it reaches a certain size at which point it is vulnerable to melting. When it melts, it does so quickly. If the ice sheet has a preferred life cycle such as this, it might help explain the 100 thousand year cycle in Earth's recent past. The idea of catastrophic ice sheet melting may also be relevant for the future (Chapters 12 and 13).

Bubbles of ancient air trapped in the cores of ice sheets reveal a conspiratorial underbelly to the orbit / ice sheet story. The bubbles have been analyzed for the concentrations of CO_2 and the other greenhouse gases. The longest ice core record so far was just published recently, extending back 800 millennia. The bubbles reveal that the atmospheric CO_2 goes up and down through the glacial cycles (Figure 8).

The similarity between CO_2 and the temperature in Antarctica is jaw-dropping. Nature simply doesn't work so cleanly most of the time. I get the impression, from reading the newspaper, that the tiniest correlations in shotgun-blasts of medical data are enough to change the diets of millions of concerned people. Are eggs good for you or bad for you this year? I can't even remember. Trans fats, are they really particularly poisonous, or are they about the same as other thick sludgey fats? Even the link between cigarette smoking and cancer, kind of a gold standard in the medical world, is not as tight as the correlation between CO_2 and Antarctic temperatures. Nonsmokers sometimes die of lung cancer, and smokers sometimes die of old age. I think it's rare

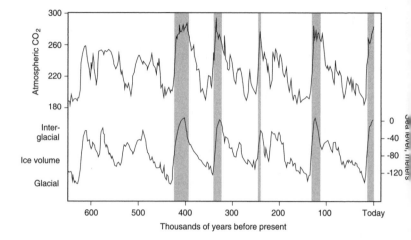

FIGURE 8. Variation of atmospheric CO_2 and sea level through the last 650,000 years. The gray bars indicate interglacial times.

to find a signal of such sublime clarity in the chaotic natural world as that between CO_2 and the temperature in Antarctica.

It is not known why or how the CO_2 concentration goes up and down through the glacial cycles. There are known factors that can account for some of the CO_2 drawdown during glacial times—changes in the ocean temperature, for example—but there are other factors that should have raised CO_2, such as the die-back of land vegetation in a glacial world. Even though the right answer is known in advance, models of the carbon cycle generally fall short of predicting the full change in the CO_2 in the atmosphere between the glacial and interglacial climate states.

The orbital variations drive the climate by allowing ice sheets to grow or causing them to melt, and CO_2 also drives climate by the greenhouse effect. The two climate drivers are about equally important to explaining the colder temperature of the glacial climate. So which is the tail and which is the dog?

Because the overall climate cycles follow orbital variations, it is tempting to assign orbital variations and ice sheets the lead role in driving the glacial interglacial climate cycles. It would be simplest to assume that the CO_2 changes somehow respond to the climate forcing of the ice sheets, acting as a kind of amplifier of the climatic dictates of the orbital variations.

The fly in the ointment here is that in the clearest climate transitions, which are the deglaciations, CO_2 goes up before the ice sheets melt. How can CO_2 be some kind of me-too amplifier if it starts to change first? It's a vexing mystery, all right. I envision the ice sheets and the CO_2 intertwined in a feedback loop of cause and effect, like two figure skaters twirling and throwing each other around on the rink. It would be very confusing to try to analyze the physics of the trajectory of either skater without considering the other.

One implication of the CO_2 cycles in the past is that the forecast of the deep future, which we will come to in Chapter 8, kind of loses credibility if we don't understand the past very well. For the near term (the coming century), the glacial cycles in atmospheric CO_2 may not be very relevant, because it took CO_2 thousands of years to change, and the last really big change—the deglaciation at the end of the last glacial period—took about 10 thousand years. Such slow processes, it might be hoped, will not rouse themselves too strongly in the coming century. On longer timescales, if the past is truly the key to the future, the carbon cycle might be expected to amplify global warming, the same way it amplified ice sheet forcing of climate in the past. We will return to this idea in Chapter 10.

CHAPTER 6
Geologic Climate Cycles

In this chapter we are going to zoom out even further in time. Over millions of years, the climate of the Earth changes in different sorts of ways, with different sorts of patterns, than we have seen so far. The climate of the last 35 million years includes ice sheets: large and permanently frozen, holding significant amounts of water. Before this time, for millions of years, there were no ice sheets at all.

A time period such as today, with permanent ice sheets somewhere on Earth, will be referred to as a "great ice age." Within our current great ice age, the ice sheets periodically grow and melt back, and global temperature rises and falls, in what are called "glacial cycles." The Earth is currently in an interglacial interval of a great ice age.

Figure 9 shows climate's descent from the hothouse world of tens of millions of years ago into the ice-age conditions of the present-day. Ice sheets existed in times indicated by the bars at the bottom of the graph. According to this reconstruction (from Zachos et al., *Science*, 2001), ice sheets in the Northern hemisphere date back less than 10 million years.

Antarctica is an ideal continent to put an ice sheet, because it is centered on the pole and surrounded by ocean all around. The opening up of the Drake Passage between South America and

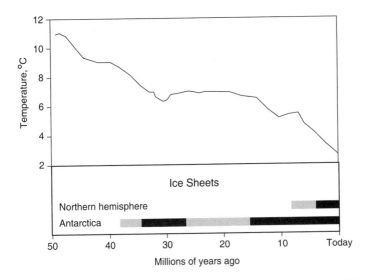

FIGURE 9. The demise of the last hot-house climate. Reconstruction of the temperature of the deep ocean through the past 100 million years. The bands at the bottom indicate strong (black) and weak (gray) evidence for ice sheets. From Pamela Martin and colleagues, Quaternary deep sea temperature histories derived from benthic foraminiferal Mg/Ca, in *Earth and Planetary Science Letters*, 198, 193–209, 2002.

Antarctica created the circumglobal moat around Antarctica, maybe playing a role in provoking the formation of the Antarctic ice sheet about 38 million years ago.

Before this time the Earth was in an ice-free "hothouse" configuration. Earth's climate was tropical all the way to the high latitudes: palm trees, crocodiles, pond algae, the works. Dinosaur dioramas at natural history museums generally look pretty tropical because this was the world most of the dinosaurs lived in.

According to the reconstruction in Figure 10, great ice ages such as our own seem to come periodically in recent Earth history. Over the past 500 million years, the Earth glaciated three times, about every 150 million years. It's not clear whether there is a predictable cycle at work here, or whether the even timing of the glaciations arose just by chance.

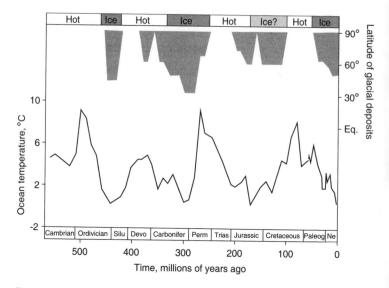

FIGURE 10. A reconstruction of Earth temperature for the past 5000 million years, showing multiple hot-house and ice age climates. From Jan Veizer and colleagues, "Evidence for decoupling of atmospheric CO_2 and global climate during the Phanerozoic eon," in *Nature*, 408, 698–701, 2000.

The transitions between hothouse and ice age climates are probably driven ultimately by processes happening deep within the Earth. Throughout the 4.6 billion year history of the Earth, the rocky material in the Earth's mantle (the layer outside of the metal core) has been turning over, like roils in a pot of water on the stove, or like a lava lamp. Warm rock rises from the depths, perhaps from the core–mantle boundary, and cooler rocks sink from the Earth's surface.

As the Earth cooled, a brittle skin formed at the Earth's surface, floating on the mantle like the skin on scalded milk. Throughout the first half of the lifetime of the Earth, called the Archean, the Earth was so hot that the plates of the Earth may not have been as rigid as they are today. Plate tectonics as it is taught in schools today may have begun only in the second half of Earth's history, about 2.5 billion years ago.

Geologists are able to reconstruct the arrangement of the continents most reliably through about the past 0.5 billion, or 500 million years. This is just the last 10% of Earth history. This is the time interval when rocks contain fossils, proper identifiable fossils of clams and trilobites with sophisticated shells and skeletal parts. The presence of fossils aids tremendously in geological reconstruction, because they can be used to assign ages to the sedimentary rocks in which they were found. Our knowledge of climates of the past is also much more reliable for the most recent 500 million years, for this reason.

Atmospheric CO_2 levels in the last hothouse climate are thought to have been higher, many times higher than today (Figure 11). Most of the carbon on Earth is tied up in rocks, with the atmosphere holding a minuscule fraction. There is no reason why the solid Earth could not spare a bit more carbon for the atmosphere. Life on Earth can handle much higher CO_2 concentrations. Humans and animals can generally tolerate CO_2 concentrations that are one hundred times higher than atmospheric, before we start getting headaches and blacking out. Plants are benefited by high CO_2 concentrations.

As with all types of geological information, our certainty of past atmospheric CO_2 concentrations deteriorates as we look deeper in time. There are no ice cores to carry bubbles of air from the ancient hothouse world; in fact, looking for ice core data from a hothouse climate is kind of a losing proposition if you stop to think about it!

However, there are traces of past atmospheric CO_2 concentrations stored in $CaCO_3$ deposits from desert soils. When water evaporates, it leaves behind $CaCO_3$ deposits, the same as those white mineral deposits you may have to scour from your bathroom fixtures. The exact chemistry of that $CaCO_3$ depends on the amount of CO_2 that was in the air when the $CaCO_3$ was formed.

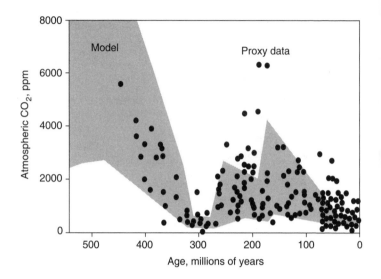

FIGURE 11. Estimates of atmospheric CO_2 concentration in the deep geologic past.

Another method of inferring past CO_2 levels involves leaves—not just tea leaves but other kinds of leaves. In the bottoms of leaves there are vents called stomata, allowing CO_2 into the inner parts of the leaf where photosynthesis occurs, making new plant material out of the CO_2. The vents also allow water to escape, which is a bad thing for the plants. This trade-off between CO_2 and water is the basis for CO_2 fertilization, which we'll discuss in Chapter 8.

Counting stomata is a way to measure past atmospheric CO_2, because plants make fewer stomata when CO_2 concentrations are high. This relationship works in the lab, and in field tests from recent times. Ginkos are good carries of stomatal CO_2 records, because ginkos have existed for 200 million years.

None of these methods is perfect and foolproof, but they generally give consistent results with each other. The atmospheric CO_2 reconstruction in Figure 11 is also consistent with climate indicators like ice sheet deposits. Hothouse climates generally have higher CO_2 concentrations than great ice age climates.

On these long, geological, million-year timescales, atmospheric CO_2 goes up and down because the solid Earth is breathing CO_2. The carbon cycle between the solid Earth and the atmosphere stabilizes the temperature of the Earth by a mechanism called the weathering thermostat. The weathering thermostat regulates the temperature of the Earth the way an electronic thermostat regulates the temperature of a heated house.

A house thermostat may take an hour to restore the temperature to its set-point. The weathering thermostat takes half a million years or longer to adjust the temperature of the Earth. The glacial cycles explored in the last chapter were faster than the response time of the weathering thermostat, and so the thermostat was not very effective at damping them out.

The weathering thermostat can be explained using the same sink analogy that helped describe planetary energy flow in Chapter 1. The water in the sink in Chapter 1 stood for heat energy on the planet, but here the water in the sink is CO_2 in the atmosphere.

Water flows into the sink, which represents CO_2 emerging from the Earth in volcanic gases and deep ocean hot springs. Water flows down the drain, which represents CO_2 uptake associated with weathering. The rate of weathering depends on the amount of CO_2 in the air, in the same way that the flow down the drain of our sink depends on the water level in the sink. The Earth regulates its atmospheric CO_2 concentration as a way to balance the CO_2 fluxes into and out of the solid Earth.

The thermostat mechanism is based on the idea that the rate of weathering, which removes CO_2, depends on how much CO_2 there is in the atmosphere. A high-CO_2 world is a tropical world, with lush rain and lots of river runoff to the ocean. If CO_2 were too low, water would freeze, and with no liquid water, weathering reactions are stopped nearly completely. A high-CO_2 world weathers and therefore inhales CO_2 from the atmosphere to the solid Earth more quickly than a low-CO_2 world does.

Long-term climate changes—such as the 100-million year timescale drift between great ice ages and hothouse climates—are caused by slow changes in the set-point of the igneous weathering thermostat. In the sink analogy, there are two ways of altering the equilibrium water level. One way is to turn up flow from the faucet, so that water runs into the sink faster. The water level would rise toward a new equilibrium. The same would happen to atmospheric CO_2 if the rate of CO_2 degassing from the Earth increased.

CO$_2$ degassing is paced by the continuous crunch of plate tectonics. Half of the CO_2 degassing from the Earth today occurs at spreading centers, where new ocean crust is formed as plates pull apart. If the rate of plate spreading was different at some time in the past, it would probably affect the rate of CO_2 degassing from the spreading centers. Like what turning up the faucet does to the water level in the sink, increasing the CO_2 degassing rate would ultimately lead to a higher CO_2 concentration in the atmosphere and a warmer climate.

When $CaCO_3$ from the sea floor is pushed down into the hot interior of the Earth (subducted), it tends to release its CO_2 to return to the atmosphere through some nearby volcano. The rate of $CaCO_3$ subduction probably varies quite a bit through time. Today, most of the $CaCO_3$ accumulation on the sea floor takes place in the Atlantic, because of the way the water circulates through the world's deep oceans. But most of the places where ocean crust subducts are in the Pacific. So the global rate of $CaCO_3$ subduction, and CO_2 emission from volcanoes, is lower than it could be.

The last time that a large $CaCO_3$-bearing piece of ocean crust subducted was when a large, shallow, tropical seaway called the Tethys Sea got caught in the collision between the Indian subcontinent and Asia. As the sinking $CaCO_3$ heated up, it may have released CO_2 to the atmosphere, raising the steady-state CO_2 concentration of the atmosphere and creating the Eocene hothouse climate.

The other way to change the water level in the sink is to alter the drain. The drain in this version of the sink analogy is weathering, and there are lots of things that can affect the rate of weathering. Soils impede weathering, by isolating the igneous bedrock from exposure to fresh rain water. Mountain-building, on the other hand, exposes bedrock to the atmosphere, accelerating the rate of weathering. High mountains tend to erode, grinding the rocks down into pebbles and sands, and in extreme cases into glacial flour, which weathers extremely quickly. The uplift of the Himalayas, driven by the ongoing collision between the Indian subcontinent with Asia, has been blamed for a cooling trend over the past tens of millions of years—the reason that the Earth is in the midst of a great ice age today.

The evolution of land plants, 450 million years ago, had a profound effect on atmospheric CO_2 by accelerating the rate of weathering. Plants take up CO_2 from the atmosphere through their leaves. Much of this CO_2 is ultimately released again down in the soil gas. The CO_2 concentration in soil air is ten times that of the atmosphere, accelerating the weathering reaction. The invasion of plants onto the land surface probably boosted the global weathering rate, pulling down atmospheric CO_2 until eventually there was enough planetary cooling to counteract the weathering boost from the plants.

Sea level change on geologic timescales plays out on a stage set by the crust of the Earth. The crust comes in two types, continental and oceanic. The rocks that make up the continental crust are chemically distinct from the rocks in the ocean crust or in the mantle below.

The continental crust is analogous to slag, the impurities in a vat of molten metal in a foundry that float to the top and collect there. Most of the continental crust of the Earth formed billions of years ago. Since then it has been shuffled around, split apart and glued back together. Sometimes continental crust becomes submerged in the ocean, accumulating a veneer of sedimentary rocks.

Ocean crust is chemically more akin to the mantle, and it sinks freely back into the Earth at places where plates converge and collide. The average lifetime of ocean crust is only about 150 million years. Continental crust, in contrast, is probably more likely to be eroded, that is to say weathered or ground down, than it is to sink back into the Earth.

Both forms of crust float in the fluid of the mantle like icebergs floating in water. Archimedes discovered that a floating body sinks to the level at which it displaces its own weight in the fluid. Ocean crust is thinner and denser than continental crust, so it floats lower than continental crust does. The waters of the ocean simply fill in the deep parts, and so it mostly covers the oceanic type of crust.

When an ice sheet grows, its weight tends to depress the crust underneath it, like driving a truck onto a ferry and watching the ferry sink down a few inches into the water. The West Antarctic Ice Sheet, described in Chapter 12, touches ground entirely below sea level. It originally formed above sea level, but then the land sank. When an ice sheet melts, the land surface beneath it rises up like grass springing back after a footstep. It takes tens of thousands of years for the crust to float up or sink down, so the landscape in Hudson's Bay, Canada under the former Laurentide ice sheet is still rising, 10,000 years after the ice melted.

You have now read about three reasons why sea level might change, (1) by adding water to the oceans from ice sheets,

(2) by thermal expansion of the water in the ocean, and (3) because of slow changes in the elevations of the plates of the Earth. On timescales of millions of years, the average height of sea level, relative to the average height of the continental crust, fluctuates for reason (4) that geologists still argue about. Maybe the solid Earth inhales and exhales water, the way it does CO_2.

The highest sea levels of the past 500 million years date to the Cretaceous period, 100 million years ago. Reconstructions from around the world tend to agree that sea level was about 250–300 meters higher than today, high enough to submerge about a third of the present-day land surface. The Cretaceous was a time when there was very little if any permanent ice on land, but if you melted all the ice today you'd get 70 meters of sea level, not 250 or 300 m. The rest of the missing sea level change is what geologists disagree over; it might have come from changes in the shape of the ocean basin, swollen mid-ocean ridges for example, or from release of water that is now dissolved in the mantle.

There is an analog for the global warming future, a trial run from the geologic past, 55 million years ago, called the Paleocene Eocene thermal maximum event, or PETM (Figure 12). The story is told in the isotopes of oxygen and carbon preserved in $CaCO_3$ in deep sea sediments and in ancient soils on land.

The carbon isotopes tell of a massive release of isotopically light carbon, probably derived from biological carbon in some form, just like our release of fossil biological carbon today. It is impossible to know exactly how much carbon was released without knowing the isotopic composition of the added carbon. Three independent ways of estimating the amount of CO_2 released do not agree completely, but the most ecumenical answer is that the magnitude of the release is comparable to the amount of coal reserves on Earth, about 5000 billion metric tons of carbon, or Gton C.

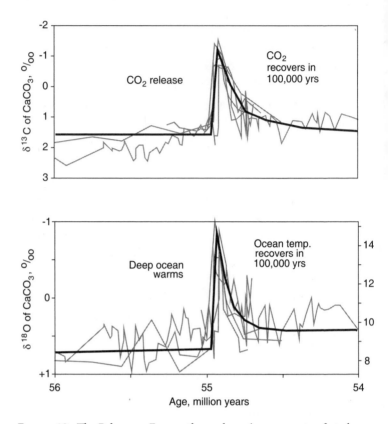

FIGURE 12. The Paleocene Eocene thermal maximum event, a fast change in atmospheric CO_2 and ocean temperature, and a slower parallel recovery of both.

The duration of the CO_2-release interval is also not known very well. It could have been nearly instantaneous, or it could have taken place over a period of ten thousand years. For comparison, the fossil fuel era will probably be all over in at most a few centuries when the coal runs out.

Along with the spike in carbon isotopes, there was a parallel spike in oxygen isotopes that tells us of a dramatic warming in the deep sea. Before the event began, the ocean was already warmer than today by about 4°C, but during the PETM the ocean warmed another 5–8°C. The warm time coincides with the time when the carbon isotopes tell us that atmospheric CO_2 was still elevated.

The CO_2 release triggered the dissolution of $CaCO_3$ from sediments in the ocean. CO_2 is an acid when it dissolves in water, and $CaCO_3$ is a solid form of a base. They react together as acids and bases are wont to do, dissolving the $CaCO_3$ in the process. The same neutralization reaction will happen in the coming millennia, as we will describe in a coming chapter (Chapter 9).

In the aftermath, both the carbon and the oxygen isotope records return to their pre-event values, on a timescale of perhaps 140 thousand years. It takes about this much time for the igneous weathering thermostat, described earlier in the chapter, to stabilize the CO_2 concentration of the atmosphere, just as it takes time for the sink to reach its steady water level. The ultimate recovery from fossil fuel CO_2 release will also take hundreds of millennia, for this reason (Chapter 9).

The PETM demonstrates that the thermostat works in 100 millennia, which is 0.1 million years. Great ice ages come and go every 150 million years, 1500 times slower than this. The water level in the sink is changing much more slowly than it could change. Instead of a single, sudden change in the configuration of the sink, a piece of carrot suddenly partially plugging the drain for example, the cause must be some slow change in the drain, say a gradual clog in the pipes, to cause a slow evolution of the equilibrium water level in the sink. The slow changes in the carbon cycle probably derive from the ponderous timescales of continental drift.

The bottom line for our forecast of the future is that the Earth has the ability to look after its own climate, but only if we are willing to wait a few hundred thousand years. It takes that long for the imbalance between CO_2 release and uptake back into the

Earth to affect the CO_2 concentration of the atmosphere and ocean. The slow response time of Earth's thermostat is the reason why our own climate experiment from releasing fossil fuel CO_2 will persist for hundreds of thousands of years into the future. This is the central topic of Section 3 of this book.

CHAPTER 7

The Present in the Bosom of the Past

The first question many people ask about the issue of human-induced climate change is how the forecast stacks up against natural variability and cycles in climate. Is global warming something big, or is it just nature-as-usual for the Earth? That question can now be answered by comparing the global warming forecast from Chapter 3 with past climate variations described in Chapters 4, 5, and 6. Even if you glazed over some of the marvelous details in those chapters, you should be able to jump back in here, reading this chapter as a sort of summary.

The impression I have is that global warming so far has been comparable to climate changes over the past millennium such as the Little Ice Age and the Medieval Optimum. In both cases, there have been regions where the climate changes are or were noticeable and harmful, but globally the impacts of the changes were subtle.

The potential climate change in the future is not subtle. If humankind burns all of the coal, the new climate of the Earth could be the warmest in tens of millions of years, since long before our evolution as a species. The transition from the natural climate to the new one could be the most severe global change since the Cretaceous / Tertiary boundary 65 million years ago that brought a close to the 150-million-year era of the dinosaurs.

The warming we have seen to date may be just noticeable to the man in the street, but it has certainly not been obvious for most of us. I've lived most of my life in the Midwestern United States, and I haven't really noticed a warming trend over my lifetime, based on my own personal experience. I moved up to Chicago from central Indiana at some point, and that probably led to more climate change for me than the global change in temperature did. Summers in Bloomington got plenty hot decades ago, I'll tell you what. The warming trend of the average Earth over the past decades is clearly measurable and it's real. I'm not trying to belittle it. It is important as an indicator of future warming, and as a test for the climate models. But in the larger picture, the impacts of that warming have been subtle.

There have been regions of the Earth that have been harder hit, for example permafrosts in high latitudes and low-lying areas affected by sea level rise. Perhaps the strongest impact of climate change to most of us so far is the frequency of extreme climate events such as strong storms, rainfall, and heat waves. But overall, my subjective impression, maybe you don't agree, is that the 0.7°C of warming that we have seen in our lifetimes has been measurable, and measurably harmful in places, but not really noticeable to most of us in our daily lives.

Climate change of the past few decades has been comparable to natural climate variations over the past millennium. A few centuries ago, from 1300 to 1860 A.D., the climate of Europe in particular was cooler than today in a climate interval called the Little Ice Age (Chapter 4). The extent of global cooling is estimated to have been about 1°C (relative to a defined natural climate of 1950). This is a bit more temperature change than we have yet experienced.

The Little Ice Age was a time of noticeably different climate to the guy on the street, not so much because of the cooling but because the character of the climate was different then, more severe. There were decades of intense cold, alternating with decades of drought, or floods. Other decades were warm and mild. Climate changed its character in the Little Ice Age more than the global average 1°C of cooling would have suggested.

Before this time was a period of general warmth called the Medieval Optimum, lasting from 12 to 8 centuries ago, 800 to 1300 A.D. Like the Little Ice Age, it is difficult to say whether the warming was global, or mostly confined to the North Atlantic region. The degree of warming is generally estimated to be somewhat less than the cooling of the Little Ice Age, perhaps 0.5°C warmer than the "natural" 1950 temperature. This is comparable to the warming we have experienced so far. For Europe it was a time of plenty, with bountiful harvests and a thriving human population. For the North American southwest, it was a time of prolonged drought.

In general, the climate changes of the past thousand years were comparable to those of the last fifty years. Temperature changes in the last millennium were noticeable but not catastrophic to most people on the Earth, although a few regions were hit especially hard. One difference is that there is no strong indication that the temperature changes in the last thousand years were global, the way the warming since 1950 has been. Natural temperature changes in the last thousand years tended to be regional, mostly averaging out on the global scale. Present-day warming is something new in that it is happening nearly everywhere, all at once.

Let's look now to the warming expected in the coming century, forecast to be 2–4°C by the year 2100. The development of agriculture and social civilization has taken place entirely in the "long summer" of the Holocene, with no climate changes as large as this.

The projected warming to 2100 is perhaps comparable to the change in temperature at the end of the last ice age, a warming of about 5–6°C. This was a very big deal; the glacial world was a different world than ours. The northern continents at that time were tundra, only inhabitable occasionally, when conditions relented a bit. The landscape of Northern Europe looked like Newfoundland or the uninhabited Norwegian island of Svalbard. When that climate ended, it changed the landscape of the Earth in ways that would have transformed life on Earth in the eyes of the human civilization that witnessed them. Today, human civilization is arguably living at or exceeding the sustainable carrying capacity of the planet, a bad time to consider large-scale transformative rearrangements of the climate.

The warm world would be something new, a climate such as the Earth has not seen in millions of years. The CO_2 record from the ice cores shows that CO_2 levels are already higher than they have been in over half a million years. Temperatures recorded in deep sediments show the present cool climate to be tens of millions of years old. The global warming climate could begin to resemble hothouse climates of the past, such as the Eocene Optimum 40 million years ago.

The problem with a novel climate is that it is unpredictable. The Earth in recent times (the last few million years) has been cooler than today, but not much warmer. Our picture of the Eocene Optimum climate 40 million years ago is really very sketchy. The last time that the climate changed as quickly as it is forecast to do was the Paleocene Eocene thermal maximum event, 55 million years ago, the details of which are even sketchier.

Humankind no doubt welcomed the end of the ice age. There is no reason I know of to say that the Eocene hothouse was an inhospitable world for living things on Earth. On the other hand, the human species has a range of optimum temperature, at which we are most comfortable and productive and healthy. It would be interesting to put it to a vote, a one-person one-vote global referendum, on whether we would prefer the Earth to be warmer, or cooler, if we had to choose. There might be some Canadians and other temperate types who might benefit from a warming, but think of all the people in the tropics. My guess is, just by force of numbers, warming would lose this vote.

A climate change of the magnitude of the deglaciation, 5–6°C, would be catastrophic to human civilization. The forecast for future warming, 3–5°C, is less than that for deglaciation, but the warming would take the planet to a climate unlike any in millions of years. A climate shift of this magnitude would rearrange the landscape and societies of the Earth.

The IPCC forecast for climate change in the coming century is for a generally smooth increase in temperature, changes in rainfall, sea level, and so forth. However, actual climate changes in the past have tended to be abrupt. The forecast resembles a simple climate response to our smoothly dialing up the CO_2, while the past looks like a series of flip-flops from one climate state to another within a few years. The forecast is based on climate models, which are for the most part unable to simulate the flip-flops in the past climate record very well, either. In this light, the forecast is a best-case scenario, because it avoids unexpected surprises (Chapter 4).

Human civilization arose during a time interval, the Holocene, that was the most stable climate interval in 650 millennia. When a complex, ordered society suddenly finds itself with more people than the landscape can support, because of prolonged drought for example, the society tends to fall apart catastrophically (see the recent book *Collapse*, by Jared Diamond). The impact of a severe sudden rearrangement of the landscape of the Earth on our society today is frightening to imagine, all the more so if it happened within a few years, rather than over the course of a few centuries.

The Paleocene Eocene thermal maximum event (PETM, 55 million years ago, Figure 12) is comparable to the potential for global warming. The PETM is of only limited use in forecasting the future, because the PETM took place so far in the past (all the way back in Chapter 6) that our picture of what happened is rather fuzzy. Life went on, of course, but there were extinctions and ultimately a whole new class of animals arose, the hoofed mammals. Hardest hit were the oceanic $CaCO_3$-forming organisms, perhaps as a result of the acidification of the ocean (Chapter 9).

It is unclear how much carbon was released in the PETM, but typical guesses are similar to the amount of fossil fuel coal available for us to burn. The deep ocean warmed by 5–8°C, and probably the land surface warmed about that same amount, too. This is considerably more than the warming projected for the year 2100, for two reasons. One is that the climate of the Earth takes a few centuries to warm. If it warms 3°C by 2100, there could be another 2°C of warming "in the pipeline," based on the CO_2 concentration of the atmosphere at 2100. Also, all of the coal will probably not be burned by the year 2100.

It's not clear how long it took for the CO_2 to be released in the PETM. It could have been faster or slower than the fossil fuel era will last. The timescale of the CO_2 release is important because if CO_2 is released faster than it can invade the ocean, the concentration in the atmosphere will spike for a few centuries until the ocean catches up. This is our situation. There is no evidence of a century-timescale climate spike at the PETM, only the long-term aftermath for 140 millennia. The PETM serves as a warning that the natural climate may take hundreds of millennia to fully recover, a topic we will address more fully in Section 3.

A little further back in time we find the Cretaceous / Tertiary boundary, 65 million years ago. This is a major turning point in the history of life on Earth—the time of the extinction of the dinosaurs, leaving room for the rise of mammals. Various individual species of dinosaurs came and went, but the age of dinosaurs lasted over 150 million years.

It is now fairly clear that the end of the age of dinosaurs was precipitated by the impact of an asteroid or comet 10 km or so in diameter, which fell to earth in what is now the Yucatan peninsula. The asteroid put so much dust in the upper atmosphere that, at the surface of the Earth, it seemed like night for months on end. Darkness killed photosynthesis, the primary source of energy production in the food web. The actual climate impact of the asteroid strike was relatively small and short-lived. The K/T event was biological, rather than fundamentally climatic.

Mankind does have the capacity to reproduce the severity of this event, but not by fossil fuel CO_2 release. It would take a substantial exchange of nuclear weapons, and the nuclear winter effect, to generate an event comparable to the Cretaceous / Tertiary boundary.

The bottom line is that the global warming climate event is not unprecedented in Earth history. Climate changes through the glacial cycles were probably as severe as global warming has the potential to be. The Earth and the biosphere will survive.

Viewed in the same time perspective, however, human civilization is also totally unprecedented in Earth history. Culture, a sort of adaptable layer of software over the hardware of animal instinct, arose about 40 thousand years ago, in the depths of the glacial climate. Human achievement and population have been growing exponentially ever since. The Younger Dryas was the largest climate change in this interval, and it proved to be the mother of the invention of agriculture.

The only interruption in the "long summer" that followed was the 8.2k event, 8200 years ago. In comparison with the climate flip-flops during the glacial time, the Dansgaard–Oeschger and the Heinrich events, the 8.2k event was pretty tame, and it lasted only a few centuries. Even though it was relatively small, the cold and drought of the 8.2k event would have been devastating to the human civilization that has arisen since. The temperature change from global warming could be much larger than 8.2k. Civilized humanity has never seen a climate change as severe as global warming.

In comparison with past climate records, the global warming forecast looks unnaturally smooth. The glacial climate seemed to flip suddenly from one stable state to another, even though the factors that drove climate to change (the orbit of the Earth, the atmospheric CO_2 concentration, and the size of the ice sheets) varied slowly. Today, the atmospheric CO_2 concentration is changing faster than it ever has in the past, as far as we know from ice core and sediment records. Yet the IPCC forecasts for the coming century look a lot less bumpy than climate records of the past do. Models of the Earth system seem less tippy than the real world, perhaps because of feedbacks, such as between vegetation and climate.

The future differs from the recent past as the Earth warms into a hothouse climate state that has not existed for millions of years. The most recent instance of a fast warming into a hothouse climate was 55 million years ago, the Paleocene Eocene thermal maximum event. Our examples of warm climates are so remote in the past that we don't know much about them. This makes it extremely difficult to predict and prepare for the future.

Section 3

THE FUTURE

CHAPTER 8

The Fate of Fossil Fuel CO$_2$

We would never have imagined life on Earth if we hadn't seen it for ourselves. The intrepid heroes on the TV show Star Trek occasionally encountered sentient beings composed entirely of energy, rather than carbon. Such beings would never have predicted the magic of carbon on Earth from first principles, or at least from the first principles of science that we have discovered so far.

A tiny fraction of the carbon on Earth is living carbon. If the living carbon on Earth were smeared out over the entire surface of the Earth (a grisly thought) it would be just a few millimeters thick. This thin layer of goo is able to accelerate the chemical reactions on the planet to rates thousands of times faster than they would go otherwise. It controls the chemistry and the climate of the surface of the planet, its atmosphere, oceans, and soils. It aggressively seeks out new chemical reactions to exploit, and has figured out how to harvest light energy from the star the planet is orbiting. Who would have thought of this?

Life is based on the chemistry of the element carbon. No other element rivals it in its complexity on Earth. Carbon's nearest relative on the periodic table, silicon, has a complicated chemistry on the Earth, too. Silicon chemistry sets the stage for plate tectonics, and the properties of the ocean and continental crusts.

The formation of soils as a product of weathering reactions is a result of silicon chemistry.

Silicon controls the Earth's interior, but carbon has claimed the surface. The carbon cycle extends into the oceans, and into solid parts of the Earth. Trees and grasses are made of carbon, and they leave carbon residues in soils. The ocean contains a thin soup of biological carbon, and much larger amounts of abiological, oxidized carbon in forms like bicarbonate ion, HCO_3^-. Carbon remnants of dead plankton, from biological tissue and as $CaCO_3$ shells, sink to the sea floor and accumulate in the sediments. Some of that carbon is carried into the deep hot interior of the Earth where the ocean plates converge in subduction zones. Perhaps the largest reservoir of carbon is the sedimentary rocks on continents, deposited there during times of high sea level or tossed from the ocean onto continental crust by the slow train wrecks of plate tectonics.

The atmosphere is one of the smaller reservoirs of carbon on Earth. If the CO_2 in the atmosphere were to freeze out into dry ice and fall to the ground in a uniform snowfall around the world, the CO_2 snow would only be about 10 cm deep. The large carbon reservoirs in the ocean, on land, and in the rocks, all exchange carbon with the atmosphere. They all breathe, like lungs of different sizes and different breathing rates. The atmosphere is Grand Central Station, a CO_2 conduit, shared by all of the CO_2 lungs on the planet.

The carbon in fossil fuels has been sleeping in its geological beds for a long time. As it moves into the atmosphere, it will provoke responses from the other parts of the carbon cycle. In some cases the response will be to take up CO_2, for example when excess CO_2 dissolves in the ocean as described in this chapter. Other carbon reservoirs may tend to release CO_2, as in the case of melting hydrates (Chapter 10), amplifying the climate forcing from the original fossil fuel CO_2.

Humanity is releasing CO$_2$ to the atmosphere, primarily from fossil fuel combustion, at a rate of about 8.5 billion metric tons of carbon per year. A billion metric tons is called a gigaton and abbreviated Gton. Seven gigatons of carbon is equivalent to 1% of the biomass of the Earth (mostly trees). Our carbon emission outweighs humanity, all the bodies of all the souls on Earth, by a factor of about twenty.

The rate of fossil fuel CO$_2$ release is tiny compared with natural exchanges of carbon between the ocean and the atmosphere. The rates of exchange between the atmosphere and the land biosphere, and between the atmosphere and the ocean, are about 100 Gton C per year, 20 times higher than the rate of fossil fuel CO$_2$ release. It may seem reassuring that humankind is releasing CO$_2$ more slowly than these natural CO$_2$ fluxes. But the fossil fuel CO$_2$ is special because it is essentially new to the fast carbon cycle of the atmosphere, ocean, and land surface. Fossil carbon, slumbering for millions of years, is being injected into the atmosphere, the Grand Central Station of the carbon cycle.

The land biosphere exchanges carbon with the atmosphere. The biosphere on land draws carbon from the air to produce leaves and new branches in the summertime, only to release the carbon again in winter. You can witness the breathing of the terrestrial biosphere in the seasonal cycle of atmospheric CO$_2$ concentration (Figure 13). The northern and southern hemispheres are out of sync with each other, because the seasons are opposite each other across the equator. There is more land in the Northern hemisphere, so the northern hemisphere biosphere breathes more deeply.

The visible carbon—the trees, elephants, people and so on— is made up of about 500 Gtons of carbon, larger than the amount of fossil carbon released so far (300 Gton C), but tiny compared to the total potential amount of fossil carbon available (5000 Gton C). The soil carbon reservoir is larger, about 2000 Gton C,

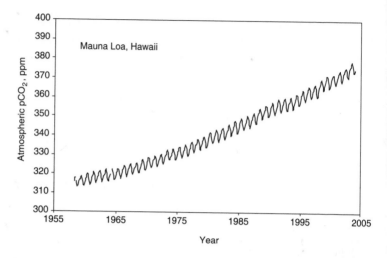

FIGURE 13. Atmospheric CO_2 concentrations of the past fifty years measured on top of Moana Loa on the big island of Hawai'i. The wiggles are the annual breathing of the biosphere, and the upward trend is caused by human emissions.

about three times the mass of the living carbon on land but still smaller than the fossil fuel carbon.

The factors that control the amount of carbon on land are complex enough that the land could serve either as a source or a sink of fossil fuel CO_2 in the coming century; it is impossible to predict which. Deforestation is already contributing CO_2 to the atmosphere, because when trees are cut down, their carbon is ultimately released to the atmosphere when the wood is burned or decomposes. Today this is mostly a tropical phenomenon, because most of the forests in temperate latitudes have already been cut, and may even be regrowing. The rate of carbon release by deforestation is about 2 Gton C per year, a bit less than a third of the fossil fuel CO_2 release rate.

There appears to be a net uptake of another 2 Gton C per year into the land biosphere, in places that are not being deforested. For a long time, this huge uptake of carbon was called "the missing sink" because no one was quite sure where the carbon was going. It is difficult to measure the amount of carbon on land, because its distribution, mostly determined by its concentration in soils, is very patchy, so it would take a huge number of measurements to tally the total accurately. Also, unlike trees and elephants, underground carbon cannot be seen with the naked eye, it must be laboriously measured. Even though 2 Gton C per year is an enormous flux, more than ten times the mass of the population of the Earth every year, our ability to see the carbon in the landscape is not good enough to find it.

The best way to determine the fate of the missing carbon is to measure the CO_2 concentration in the atmosphere, rather than in soils. To simplify the story for the sake of explanation, let's imagine that the wind blew only from west to east, and that the CO_2 concentration in the air just above the land surface decreased steadily from west to east. The atmospheric CO_2 measurements would imply that CO_2 is invading the land surface as the wind blows through.

In the real world, the winds are not steady but blow in all different directions in the chaotic turbulence of weather. CO_2 measurements are made not just at a single pair of locations, once and for all, but every day, from a global network of dozens of locations around the globe. The differences in CO_2 concentration around the globe are pretty tiny, and the results of different laboratories have to be calibrated carefully. The CO_2 data are analyzed using computer models of the winds, the same sorts of models that predict the weather. These studies commonly conclude that the missing CO_2 has gone to ground in the high Northern latitudes, the great northern forests of Canada and Eurasia.

It is not clear where exactly the carbon is going, so neither is it known why it's going there. It could be that the land takes up new carbon because the growing season is longer in a warming world. Changes in the length of the growing season have been clearly documented in agricultural records. Warming is most intense in high latitudes, explaining why the high latitudes appear to be taking up carbon now.

Alternatively, the land could be taking up carbon because of nitrogen deposition, a by-product of internal combustion engines. At high temperatures in the engines' power cylinders, nitrogen gas from the atmosphere is converted into a family of nitrogen–oxygen compounds called NOx. These compounds contribute to ozone formation in urban polluted air. The NOx compounds then degrade to nitric acid and rain out, comprising about a third of acid rain. (The rest is sulfuric acid.) The nitrogen then winds up in soils in a chemical form called nitrate (NO_3^-), which is a plant fertilizer. An increase in the deposition of nitric acid rain might be fertilizing plants to take up the extra carbon.

A third possibility is that higher CO_2 concentration itself could be fertilizing the plants. Plants grow better, everything else being equal, in higher CO_2 air. Plant growth requires CO_2, just as it requires fertilizers like nitrate. CO_2 is obtained through vents called stomata in the air-tight waxy seals around leaves. The waxy leaf walls exist to prevent loss of water vapor. Opening the stomata in order to inhale CO_2 entails a certain unavoidable loss of water. Higher CO_2 concentration in the air enables the leaf to get the CO_2 it needs without opening the stomata as much.

CO_2 fertilization goes only so far toward stimulating plant growth, however, because plant growth is typically more closely limited by fertilizers like nitrate, rather than by CO_2. Forest scientists have made measurements of the CO_2 fertilization effect by releasing CO_2 into the air upwind of a grove of trees, and then measuring the growth rates relative to the rates of "control" trees that have no extra CO$_2$. Imagine steel towers in a ring like Stonehenge, blowing CO$_2$ on a grove of trees from the upwind direction, 24 hours a day, for years on end. Typically, the trees grow faster for a few years, but then the boost wears off, and the growth rates taper off to normal.

The land surface today just about balances out as a carbon source or sink. Deforestation is almost balanced by high-latitude uptake, formerly known as the missing sink. It is difficult to predict what will happen to the land carbon in the coming century, whether it will be a source or a sink. High latitude uptake of CO$_2$ could continue, or it could taper off as the CO$_2$ fertilization effect saturates (if that is the reason for the present-day uptake). Or the land could start giving off excess carbon, as the bacteria and fungi that decompose organic carbon in soils do their work more quickly as temperatures rise. Tropical soils do not have as much organic carbon as higher latitude soils do, so a transition to a tropical world might in the end reduce the amount of carbon stored in the landscape.

Decisions that people make about how to use the land will have an enormous impact on the land's ability to store carbon. It is possible to increase the inventory of carbon in agricultural soils by practices such as no-till farming and crop rotation. Of course people also think of planting trees as a solution to global warming. A tree obtains its carbon from the atmosphere, storing it in its woody trunk and branches. In order to serve as a persistent carbon sink, however, the land has to remain forested indefinitely. It makes no sense to cut a forest and then claim carbon storage credit for letting the forest grow back, or to claim the carbon credit now and cut the forest next year.

It is easy to imagine the land carbon pool taking up or releasing a few hundred Gton C. But it is more difficult to imagine the land taking thousands of Gton C, even if fossil fuel carbon release approaches the 5000 Gton C of the coal deposits. To do so would mean doubling or tripling its carbon inventory, which is today about 2000 Gton C. Ultimately, carbon uptake by the land biosphere runs out of lung capacity.

The oceans are a much larger carbon reservoir. Svante Arrhenius, the first climate modeler to predict the climate sensitivity a century ago (Chapter 1), was not concerned about anthropogenic global warming. He recognized that mankind was "evaporating our coal mines" into the atmosphere, but estimated that it would take a millennium for human industry to double the CO_2 concentration of the atmosphere. Based on the CO_2 emission rates of his day, his was a reasonable and conservative projection. However, rates of CO_2 emission have gone up exponentially since then. Our current plight would have seemed like wild-eyed alarmism, had Arrhenius predicted it in 1896.

Scientists in Arrhenius' day also assumed that the oceans would take up any extra CO_2 quickly. The oceans cover 70% of the Earth's surface. The ocean is physically thinner than the atmosphere, 4 kilometers compared to the scale height of the atmosphere, which is about 8 kilometers. Water also flows around a lot, at least more than the land surface does. It makes sense to naively expect that the oceans would interact with the atmosphere pretty quickly.

It turns out though that the amount of time it takes for the atmosphere and the ocean to exchange CO_2 is measured in centuries. Most of the surface of the ocean is warm, while most of the water in the abyss is very cold. The warm water is prevented from mixing with the colder water underneath because the cold water is denser.

Imagine you are standing on the deck of an oceanographic research ship in tropical waters, say near Tahiti. Although the air and surface water are balmy, the water a few kilometers under your feet is near freezing. If you go for a snorkel, the last thing you need worry about is suddenly swimming through a patch of near-freezing water recently mixed up from the deep ocean. If the air were that cold a few kilometers away, you can be sure that you'd get a blast of it now and again. But the cold water simply can't get up here from down there.

The cold waters in the ocean abyss are exposed to the atmosphere only in high latitudes, in the North Atlantic and around Antarctica. These regions do not cover 70% of the Earth's surface, but only maybe 2 or 3%. The invasion of fossil fuel CO$_2$ into the deep ocean has to pass through this very tiny area. Fossil fuel CO$_2$ needs to dissolve into a parcel of seawater that is cold enough or salty enough—in other words, dense enough—to sink into the abyss. Water this dense is created only sporadically, in the coldest weather. Often it is made under sea ice, out of reach of the atmosphere. The bottom line is that it takes a long time, measured in centuries, for CO$_2$ to make its way into the deep ocean (Figure 14).

It is possible that with changes in surface climate, the circulation of the ocean could stagnate for a few centuries, slowing the CO$_2$ invasion into the oceans. Continued flow of surface waters into the deep sea requires water at the sea surface that is as dense as used to be there in the past. If Greenland were to melt within a century, it would probably stop the formation of deep water in the North Atlantic. If surface waters warm, the density of the surface waters will decrease. There is also projected to be an increase in rain and snowfall in high latitudes, associated with the warming. If surface waters freshen, the density of surface waters would decrease still further. Future ocean stagnation could also slow the flow of heat into the ocean, affecting the rates of climate change and sea level rise (Chapter 12).

In the past decades, the oceans have been taking up fossil fuel carbon at a rate of about 2 Gton C per year, about one-third of the rate of fossil fuel CO$_2$ release. The atmosphere today contains about 200 Gton C more than the natural atmosphere contained

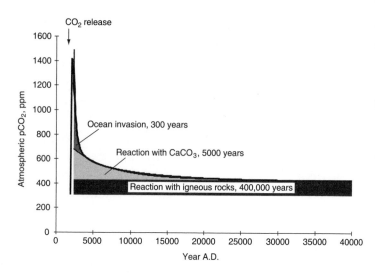

FIGURE 14. Model simulation of the CO_2 concentration of the atmosphere for 40,000 years following a large CO_2 release from combustion of fossil fuels. Different fractions of the released CO_2 recover on different time scales. From Archer, 2005.

back in the year 1750. The oceans are therefore absorbing about 2 out of 200 or 1% of the excess CO_2 in the atmosphere each year. If CO_2 uptake continued at this rate, and if we stopped releasing more CO_2, then it would take about a century for atmospheric CO_2 to return to its natural value.

But the invasion of CO_2 changes the chemistry of the ocean, decreasing its ability to hold more CO_2. When CO_2 dissolves in seawater, it reacts with a molecule called carbonate ion, $CO_3^=$. The two minus signs signify that the carbonate ion has an electrical charge of minus two. The product of the reaction is bicarbonate, HCO_3^-. The reaction is written as

$$CO_2 + CO_3^= + H_2O \leftrightarrow 2\ HCO_3^-$$

The end result is to convert CO_2, a dissolved gas that can evaporate into the air, into bicarbonate, a charged molecule that cannot evaporate. Charged molecules are what salt is made of. Try boiling away a pan of salt water, and what you will be left with eventually is a dry pan of salt.

The CO_2 gas hides in seawater as bicarbonate, invisible to the atmosphere, enabling the seawater to hold a lot more CO_2 than it would if there were no carbonate ion for CO_2 to react with. This chemical system is called a buffer. Our blood plasma uses the same chemistry to aid in the transport of CO_2 from our muscles to our lungs for exhaling. Oxygen has no such buffer chemistry, so animals with circulation systems like ours require hemoglobin molecules to carry oxygen in their blood. It is lucky for us that the concentration of $CO_3^=$ in the ocean is as high as it is, because without the $CO_3^=$ buffer, the ability of the oceans to take up fossil fuel CO_2 would be almost negligible.

The amount of CO_2 that seawater can hold is limited by the availability of carbonate ion, $CO_3^=$. As the CO_2 concentration in the air increases, seawater in contact with the air begins to run out of $CO_3^=$. The buffer loses its strength and seawater loses its extra capacity to absorb CO_2. Ocean uptake of CO_2 slows down, to a timescale of several centuries instead of the naive single-century estimate above.

More importantly in the long term, the decreasing capacity of the buffer decreases the total amount of fossil fuel CO_2 the ocean will ultimately be willing to take up at all. If the oceans were infinitely large, or if the buffer were infinitely strong, the oceans could eventually absorb all of the released CO_2, but things being as they are, the oceans will take most of the CO_2, but not all of it.

There have been several independent modeling studies, including a few of my own, to predict the eventual distribution of a slug of fossil fuel CO_2 between the atmosphere and the ocean (Table 1). The models differ in how they are put together, and what assumptions they make about the circulation in the ocean, the role of biology in the ocean carbon cycle, and other issues. However, the models find that 20 to 40% of a slug of new CO_2 will still remain in the atmosphere after the atmosphere and

Table 1.

Model predictions of the equilibrium distribution of a slug of new CO_2 between the ocean and atmosphere, with no sediment feedbacks.

	Moderate 1000–2000 Gton C	Large 4000–5000 Gton C
CLIMBER (Archer and Brovkin, in press)	22%	34%
HAMOCC (Archer 2005)	22%	33%
Lenton et al. 2006	21–26%	34%
Goodwin et al. 2007	24–26%	40%
GENIE (Ridgwell et al. 2007)		31%

ocean have reached equilibrium. If the air in the atmosphere and the water in the ocean were contained in an unreactive Teflon flask in the laboratory, instead of in a bed of reactive rocks, then 20–40% of the CO_2 released from fossil fuel combustion would remain in the atmosphere forever.

It matters how much CO_2 is ultimately released. More CO_2 acidifies the ocean more strongly, reducing its capacity to hold CO_2. In the table, a large release is assumed to be 4000–5000 Gton C. This is based on geologists' estimates of how much minable coal there is. There are also thousands of Gton C in methane hydrates and peats, discussed in Chapter 10. A moderate CO_2 release of 1000–2000 Gton C is comparable to projected fossil fuel emissions to the year 2100 under the business-as-usual scenario. This scenario would leave considerable amounts of coal in the ground.

In the real world, the leftover fossil fuel CO_2 in the atmosphere will react with $CaCO_3$ and eventually with igneous rocks. These chemical reactions act to restore the atmospheric CO_2 concentration to its stable, natural set point. This is the topic of Chapter 9, next up.

An issue to consider for the future is the mysterious CO_2 changes associated with the glacial / interglacial CO_2 cycles. The orbit of the Earth seems to be the primary driver of the growth and collapse of the ice sheets, but somehow the carbon cycle manages to amplify the climate response to the changing orbit. Does this mean that the Earth's carbon cycle will amplify global warming, also? That question will be considered, but not answered, in Chapter 10.

The bottom line of this chapter is that the natural world takes up fossil-fuel CO$_2$ more slowly than we might have expected given how much CO$_2$ is dissolving in the oceans today. Models of the carbon cycle in the ocean show that as CO$_2$ invades the oceans, the chemistry of the oceans changes, limiting the uptake of more CO$_2$. After several centuries when the oceans have inhaled their fill, a significant fraction of the fossil fuel CO$_2$ will remain in the atmosphere, affecting the climate for millennia into the future.

CHAPTER 9

Acidifying the Ocean

CO_2 is an acid, with which humankind is acidifying the oceans. In response, $CaCO_3$ (a base) will dissolve on land and in the ocean, restoring the pH balance of the ocean. This process will take thousands of years. As the pH of the ocean recovers, the change in ocean chemistry will extract some of the fossil fuel CO_2 from the atmosphere, but even after the pH recovery is complete, models of the carbon cycle predict that about 10% of the fossil fuel CO_2 will remain in the atmosphere, until it is consumed by the weathering thermostat, which stabilizes the climate of the Earth on a timescale of hundreds of millennia, as described in Chapter 6

Acidity is a property of water-based concoctions like seawater or rainwater. An acidic solution contains a high concentration of positively charged hydrogen ions, H^+. Hydrogen ions are aggressive little guys, reacting quickly and harshly with many kinds of chemical bonds, including bonds in metals, rocks, and biological carbon compounds. Your digestive juices are acidic, to aid the breakdown of the chemical bonds in your food. Battery acid is so acidic and reactive that it is dangerous; a splash will burn you.

H^+ can be extracted from a water molecule, leaving behind a remnant called hydroxyl ion, OH^-:

$$H_2O \leftrightarrow H^+ + OH^-$$

A solution with a high concentration of OH^- is called a base. A strong base solution can be as reactive as a strong acid. Drain cleaner is an example of a strong base.

If an acid is mixed with a base, the H^+ and OH^- ions combine to form water, H_2O, in the reaction above running from right to left. This is called a neutralization reaction. Because H^+ and OH^- react together, any sort of water-based solution like seawater can have either a high H^+ concentration, or a high OH^- concentration. That is, it can be either acidic or basic, but never both at the same time.

The measure of the acidity of a water sample is its pH. A low pH solution is acidic, while high pH is basic. A pH of around 7 is neutral, an equal number of OH^- and H^+ ions. Battery acid might have a pH of 1 or less, drain cleaner dissolved in water might have a pH of 14 or higher. Natural waters range typically from about 5, a very mild acid, to 9, slightly basic. Sea water tends to be slightly basic, with a pH of 7 or 8.

The ocean becomes more acidic when CO_2 invades. The way CO_2 affects the pH of seawater is by combining with water to make carbonic acid, H_2CO_3:

$$CO_2 + H_2O \leftrightarrow H_2CO_3$$

H_2CO_3 is an acid because it releases H^+, leaving behind bicarbonate ion, HCO_3^-. The pH of rainwater, comprised of distilled water and CO_2 gas, is slightly acidic because of the acidity of the CO_2 dissolving into cloud drops from the atmosphere. Because it contains CO_2, the atmosphere has an acidic influence on natural waters.

Limestone, a type of rock with formula $CaCO_3$, is chemically a base. When $CaCO_3$ dissolves, it produces calcium and carbonate ions:

$$CaCO_3 \rightarrow Ca^{2+} + CO_3^{=}$$

The carbonate ions scrounge the solution of available H^+ ions they can find, to convert themselves to the pH-neutral form of carbon, which is bicarbonate:

$$CO_3^{=} + H^+ \rightarrow HCO_3^{-}$$

When carbonate mops up H^+ to form the bicarbonate ion (HCO_3^{-}), it neutralizes the acidity of the solution, acting like an anti-acid, another word for which is base. When you have a sour (acid) stomach, and you swallow an antacid ($CaCO_3$) to neutralize it, with luck you will feel better soon.

CO_2 dissolves in seawater by reacting with carbonate ion or $CO_3^{=}$ (Chapter 8). When the ocean is acidified, its carbonate ion gets used up, decreasing the amount of CO_2 that the ocean can further dissolve. $CaCO_3$ dissolution serves to restore the carbonate ion concentration of the ocean to something like its natural value. Restoring the carbonate ion to its natural concentration is essentially the same process as restoring the pH of the ocean to its natural value.

The process of $CaCO_3$ (limestone) neutralization of the ocean accounts for a second stage of CO_2 drawdown from the atmosphere. CO_2 invasion depletes the carbonate ion ($CO_3^{=}$), and $CaCO_3$ dissolution restores it. A neutralized ocean is able to hold more CO_2 than an acidified ocean. After neutralization is complete, carbon cycle models predict that about 10% of the fossil fuel CO_2 slug will still live in the atmosphere (Table 2).

The strongest impact of ocean acidification is to the organisms that produce shells or other biological infrastructure out of $CaCO_3$. $CaCO_3$ becomes chemically unstable if the CO_2 concentration becomes too high. Carbonate secreters were singled out for extinction in response to the acidification of the ocean during the PETM climate event 55 million years ago (Chapter 6).

Table 2

Model estimates of long-term lifetime of fossil fuel CO_2 in the atmosphere, from carbon cycle models including atmosphere, ocean, rock weathering, and sediment deposition.

	Airborne fraction of CO_2			Temperature, °C		
	Peak	1 kyr	10 kyr	Peak	1 kyr	10 kyr
1000–2000 Gton C release						
CLIMBER (Archer and Brovkin 2007)	52%	29%	14%	4.5	3.1	1.8
HAMOCC (Archer 2005)	58%	24%	11%	4.9	2.7	1.4
Lenton et al. 2006	55%	18%	11%	4.7	2.1	1.4
4000–5000 Gton C release						
CLIMBER (Archer and Brovkin 2007)	67%	57%	26%	8.4	7.8	5.2
HAMOCC (Archer 2005)	60%	33%	15%	8.0	5.9	3.7
Lenton et al. 2006	72%	15%	12%	8.6	3.7	3.2
GENIE (Ridgwell et al. 2007)	50%	34%	12%	7.4	6.0	3.2
Tyrrell et al. 2007	70%	42%	21%	8.6	6.7	4.6

It has been observed, in closed environments and in the natural world, that corals produce $CaCO_3$ more slowly as the CO_2 concentration increases. The growth of a reef is always a race against time and natural destruction. Reefs are continually under attack by organisms that bore into the $CaCO_3$ in their search for food. Waves, storms, and now human activity are also continually breaking pieces of the reef. If the production rate of building material decreases to slower than its destruction rate by borers and natural erosion, the reef fails.

Ocean acidity is not the only threat to coral reefs on Earth today. Corals are very sensitive to changes in temperature. If the temperature of the water rises above a threshold value, the corals respond by expelling their symbiotic algae guests, called zooxanthelli. The algae do photosynthesis, helping to feed the coral organism itself (which is an animal, as incapable of doing photosynthesis as we are). The zooxanthelli also provide the characteristic color of the coral and the reef. The expulsion of zooxanthelli is called "bleaching." Sometimes the algae are replaced by a different species or strain that has a different, more suitable temperature tolerance. But bleaching is a desperate remedy of last resort, quite often followed by the death of the coral.

Corals are also sensitive to the clarity of the water column. Their photosynthetic algae require light to operate, which they don't get if the water gets too cloudy. Corals do best in waters that have very low concentrations of dissolved nutrients, because nutrients feed the growth of algae suspended in the water column, which steal all the light. Corals are threatened by runoff of agricultural fertilizer, which stimulates plankton growth, and by erosion of soils, which makes the runoff waters more turbid.

Open-ocean plankton called coccolithophores produce tiny shells of $CaCO_3$. Many types of plankton armor themselves with mineral shells, perhaps to make themselves less palatable to those who would eat them. The shells of coccolithophores look like tiny paper plates, pasted on the outside of a spherical cell. The rate at which coccolithophores produce $CaCO_3$ has been shown to be sensitive to changes in the carbonate ion concentration of the seawater in which they grow.

Coccolithophores play many roles in the environment and in the carbon cycle. Their plates of $CaCO_3$ are one of the main avenues for $CaCO_3$ deposition on the floor of the deep sea. In the upper ocean, the plates scatter light in the ocean, intensely enough that large coccolith blooms can be seen from space, altering the reflectivity of the ocean and hence the energy balance of the planet. Coccolithophores produce a dissolved gas called dimethyl sulfide, which ultimately evaporates to the atmosphere and alters the drop sizes of clouds in remote marine areas.

Deposition of $CaCO_3$ on the sea floor plays a role in the neutralization of fossil fuel CO_2, as discussed later in this chapter. Sinking $CaCO_3$ may also play the role of ballast, enabling the less-dense tissues from dead plankton to sink into the deep sea.

This material is the ultimate food source for almost all of the exotic biota of the deep sea. On timescales of millions of years, the burial of plankton tissue in sediments is also the source of the oxygen in the atmosphere that we breathe.

Fish and other marine animals are sensitive to acid, if it is strong enough. Fish are sensitive to acid rain for example. An entirely different phenomenon from CO_2 acidifying the ocean, acid rain refers to the deposition of the sulfuric and nitric acid by-products of coal combustion and automobile use. If acid rain falls on a terrain of $CaCO_3$ rocks, such as where I live in the American Midwest, the $CaCO_3$ dissolves, neutralizing the acid. In landscapes without available $CaCO_3$, like the American Northeast and Scandinavia, the ground waters and streams become acidic. The acidity of the water provokes aluminum to dissolve, which is toxic to fish.

Fish are also sensitive to the acidity from rising CO_2 concentration, but CO_2 is more than just an acid in this case. It is also the product of respiration, the gas that the fish are trying to exhale through their gills just as we exhale CO_2 from our lungs. Laboratory evidence shows that fish are more sensitive to acidity changes from CO_2 than they are from the same acidity change from other acids, such as those in acid rain. In general, however, fish are not as endangered by acidifying the ocean as are the carbonate secreting organisms such as corals and coccolithophorids.

It will take several millennia for $CaCO_3$ to neutralize the fossil fuel CO_2. The neutralization will take place as part of the natural cycle of $CaCO_3$ in the ocean. Dissolved $CaCO_3$ from weathering reactions on land is carried to the ocean by rivers. $CaCO_3$ is re-formed as a solid by corals and plankton such as coccolithophorids as discussed above, and removed from the ocean as it accumulates on the sea floor. If there is more input of $CaCO_3$ to the ocean than output, then the excess dissolved $CaCO_3$, building up in the ocean, reacts with CO_2. This means that the $CaCO_3$ cycle in the ocean neutralizes fossil fuel CO_2.

The dissolution rate of $CaCO_3$ on land is limited by the availability of fresh rainwater to carry the dissolved $CaCO_3$ away. The global average change in rainfall in the IPCC forecast is to increase about 3–5%, dissolving only a small amount of extra $CaCO_3$. This means that dissolution of $CaCO_3$ on land will neutralize the acid ocean only very slowly, over thousands of years.

Another source of $CaCO_3$ for neutralizing CO_2 is in the sediments on the sea floor. The distribution of $CaCO_3$ on the sea floor resembles snow-capped mountains, with white $CaCO_3$ accumulating on the shallow edges of the ocean and on the mountain-tops of the mid-ocean ridges, and dark clay-rich sediments down in the valleys of the abyss. This pattern of $CaCO_3$ distribution was one of the first things known about the landscape of the sea floor, dating back to a 1947 global voyage of oceanographic discovery called the Swedish Deep Sea Expedition.

Millions of years from now, scientists in some future Swedish Deep Sea Expedition will mark our era in sediment cores by a clay layer, devoid of $CaCO_3$. As $CaCO_3$ dissolves, it leaves clays behind. Eventually, the surface of the sediment becomes covered with a cap of clay that slows down further $CaCO_3$ dissolution. Such clay layers have been found in sediments from the time of the Paleocene Eocene thermal maximum event, described in Chapter 6. A natural release of CO_2 at that time burned down the $CaCO_3$ sediments in the ocean, just as fossil fuel CO_2 will do in our future.

Acidifying the ocean perturbs the $CaCO_3$ cycle in the ocean, further affecting the neutralization process. $CaCO_3$ production by corals and coccolithophorids is likely to slow down in an acid ocean. If less $CaCO_3$ is produced, less will get buried, contributing to the imbalance in the $CaCO_3$ cycle. Suffering coccolithophores and corals are doing their part to help neutralize the fossil fuel CO_2. Once the $CaCO_3$ is produced at the surface ocean, it has to sink to the sea floor before it can be permanently removed from the ocean. It may be that $CaCO_3$ will dissolve more extensively in the water column, rather than sink, in a more acidic ocean. This would also contribute to imbalance in the $CaCO_3$ cycle.

The rates of $CaCO_3$ production in the ocean and accumulation on the sea floor are almost 100 times slower than the rate of fossil fuel CO_2 release. So it will take some time for the neutralization to catch up with the acidification. Carbon cycle models estimate that it will take 2–10 millennia before it is complete.

Our ocean will be acidified more intensely than typically happens naturally, because the CO_2 concentration in the atmosphere is changing a lot faster now than it usually does naturally. Atmospheric CO_2 changes through the glacial cycles were as slow as the $CaCO_3$ neutralizing timescale of millennia. CO_2 rose from the glacial value to the natural interglacial value in about 10 millennia. It is a little harder to know how long it took for the CO_2 rise that triggered the Paleocene Eocene thermal maximum event (55 Myr ago, Chapter 8), but it could have taken as long as 10 millennia. In contrast, atmospheric CO_2 today is going up on a timescale of centuries. When CO_2 is released slowly enough that $CaCO_3$ neutralization can keep up, you don't get the acid spike. Our problem is that CO_2 is rising so quickly that the natural neutralizing mechanism will be temporarily overwhelmed.

The intensity of the acid spike is unprecedented through the time span of the ice core records, over half a million years. Significant amounts of CO_2 can be released to the atmosphere by large volcanic eruptions, such as occurred in India about 65 million years ago to produce 500,000 square kilometers of volcanic rock called the Deccan traps. These eruptions took place over millions of years. I don't think there is any clear evidence for a change in atmospheric CO_2 concentration in the past that was as fast as it is changing now.

There were times in Earth's history, millions of years ago, when the natural CO_2 concentration of the atmosphere could have been 10 times higher than today. If we were to increase our CO_2 concentration by a factor of 10 all at once, the concentration of carbonate ion in the ocean would decrease by about a factor of 10, a severe acidification. However, there was still $CaCO_3$ deposition in oceans of the deep past, so the ocean must not have been all that acidic. The answer to this puzzler is that in the geologic past, atmospheric CO_2 reached those higher levels slowly, over millions of years, allowing the $CaCO_3$ cycle to keep the ocean neutralized, rather than all at once, as we are doing today.

The long atmospheric lifetime of CO_2 implies that global warming will last a long time. Estimates of how warm it could be, and for how long, from various carbon cycle model studies are summarized in Table 2. The studies come to very similar conclusions.

The warming to expect in the distant future depends on how much CO_2 is released. As we assumed in the previous chapter, a "moderate" CO_2 release might be 1000–2000 Gton C, which could be achieved by business-as-usual for the coming century but leaving some fossil fuels in the ground. A "large" release would be 4000–5000 Gton C, which is essentially all of the fossil carbon.

Global mean temperature spikes for a few centuries, perhaps by as much as 5 to 8°C depending on the amount of carbon released. This is more warming than the IPCC forecast for the year 2100, because it takes a century or two for a full climate warming response to play out. It takes that long to warm the ocean. In the forecast for the 2100, there is still excess warming "in the pipeline" which has been paid for, as excess CO_2 in the atmosphere, but not yet delivered.

The models all predict a long tail for fossil fuel CO_2 concentration in the atmosphere. The excess CO_2 in the atmosphere one thousand or ten thousand years in the future may not be the exact same CO_2 molecules that come from the coal, because CO_2 molecules are continuously exchanged between the atmosphere, the ocean, and the land biosphere. But the models show that the atmospheric load of CO_2 would be higher if CO_2 is released now than if it is not.

Some studies predict that it will take 1–2 millennia for the $CaCO_3$ neutralization to take place, while others predict a neutralization time of 5–10 millennia, but these differences do not affect the bottom line conclusion. All of the studies predict that an amount of carbon equal to about 10 or 12% of the fossil fuel CO_2 will still remain in the atmosphere after ten thousand years. This leftover CO_2 must await reaction with igneous rocks, the "weathering thermostat" described in Chapter 6.

The long lifetime of fossil fuel CO_2 in the atmosphere translates to a long lifetime of global warming. A moderate CO_2 release could keep the Earth 3°C warmer for a thousand years, while a large CO_2 release could keep things 3°C warmer than natural for ten thousand years. The Greenland ice sheet is expected to melt, ultimately, in a world 3°C warmer. If humanity burns all the coal, the warming response would last long enough to melt even the most sluggish Greenland ice sheet model (more about sluggish ice sheet models in Chapter 11).

The sequence of events, then, is this. CO_2 is released to the atmosphere. On a timescale of centuries, most of it invades the ocean, leaving 15–55% in the atmosphere. The invading CO_2 acidifies the ocean, provoking an imbalance in the $CaCO_3$ cycle, which acts to neutralize the acid. The pH chemistry of the ocean recovers on a timescale of maybe 2–10 millennia. After this time, the models predict 10% or so of the fossil fuel CO_2 to remain in the atmosphere for hundreds of millennia into the future.

Life forms that secrete $CaCO_3$, such as coral reefs, may find themselves scoured out of existence if the ocean absorbs too much acid. Corals are the poster children, but there are also microscopic algae that secrete $CaCO_3$ in the open ocean, an important component of the ocean carbon cycle. Many of these types of organisms went extinct at the Paleocene Eocene thermal maximum and the K/T boundary climate events described in Chapter 6. Our CO_2 acidity storm could be harsher those in the past, because atmospheric CO_2 concentration is increasing more quickly than it typically has in the past.

CHAPTER 10

Carbon Cycle Feedbacks

The carbon cycle as presented in Chapters 8 and 9 generally has a calming influence on climate. The ocean takes up most of the fossil fuel CO_2 in a few centuries, leaving some behind, in a fairly well-behaved, predictable way. Our only complaint was that the carbon cycle was so slow to clean up the mess.

The carbon cycle from the real world, as documented in ice cores and other climate records, seems to have had a different temperament. Instead of moderating the climate changes that were driven by orbital variations, the carbon cycle seems to fan the flames of climate change. It is likely that, given time, the carbon cycle will amplify the effects of global warming in the future, as well.

There are times in the past in which warming seemed to trigger atmospheric CO_2 to rise. The interplay between temperature and CO_2 is tricky to untangle because an increase in CO_2 drives the Earth to warm, but here I claim that a natural change in temperature may also drive CO_2 to change. Put the pieces together into a loop of cause and effect, and the result is a positive feedback that makes the climate tippier.

One example is the warming and rising CO_2 concentration at the end of the glacial time, already presented in Chapter 5 and shown in Figure 8. As the ice sheets melted, over a time period of about ten thousand years, atmospheric CO_2 rose from a glacial value of 200 ppm up to an interglacial value of about 260, ultimately to drift slowly up to 280, the value it still had a few centuries ago. By the time the transition was complete, the change in CO_2 was large enough to explain about half of the temperature difference between the glacial and the interglacial climates. The rest of the temperature change can be attributed to the bare ground, which absorbs incoming sunlight, replacing ice sheets, which used to reflect sunlight back to space (the ice albedo effect).

The temperatures rose in both Greenland and Antarctica, but not in complete synchrony with each other. The beginning of the end of the ice age was initiated by slow changes in the orbit of the Earth, leading to warmer summer sunshine in the Northern hemisphere. However, the actual warming started first in Antarctica, which then stalled while it warmed abruptly in Greenland. After a thousand years of warmth in the North, the cold of the Younger Dryas interval descended on the North, at which point warming proceeded again in Antarctica.

The warming in Antarctica began a few centuries before the CO_2 began to rise. Climate contrarians have argued that because temperature has an effect on CO_2, there is no need for CO_2 to drive temperature. This is a silly argument; there is no reason why the cause-and-effect relationship between CO_2 and temperature cannot be a two-way street. The initial Antarctic warming may have been triggered by a change in the Earth's orbit, or a shift in ocean circulation patterns. Rising CO_2 amplified the initial warming, so that most of the warming associated with the end of the ice age happened after the CO_2 started to rise.

Another example of temperature driving CO_2 is the Little Ice Age (Chapter 4). This cold interval lasted from about 1300 to 1860, and is correlated with the Maunder minimum, a time when the Sun had no sunspots. From the sunspot observations,

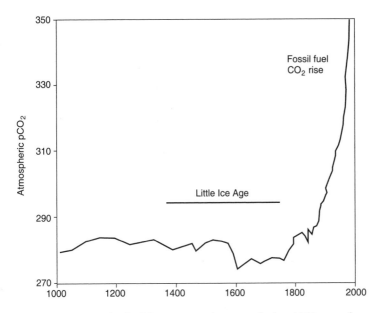

FIGURE 15. Atmospheric CO_2 concentration over the last 1000 years from the Taylor Dome ice core in Antarctica.

and from the carbon-14 and beryllium-10 proxy records of solar intensity, climate scientists surmise that a cooler Sun was the cause of the cooler Earth. There was a small decrease in atmospheric CO_2 through this time period, about 5–10 ppm (Figure 15). This change in CO_2 is enough to account for perhaps a third of the cooling associated with the Little Ice Age.

In both of these examples, the temperature change was driven in large part by external forcing, from a changing orbit at the end of the ice age, or a cooler Sun through the Little Ice Age. Atmospheric CO_2 responds to changing temperature, driving the temperature a bit further in the direction of the external forcing (warmer during the deglaciation, and cooler during the Little Ice Age). The CO_2 response acts as an amplifier of the externally driven climate forcing. These time periods are consistent with an increase in the atmospheric CO_2 concentration of 10–50 ppm per °C rise in temperature. An initial 1.0°C rise in temperature might change CO_2 levels enough to raise temperature another 0.1 to 0.7°C.

The most likely source of a carbon amplifier in the coming century is probably the terrestrial landscape, the trees and soils. The landscape today is a net wash as a carbon source or sink. Carbon release by tropical deforestation is just about compensated by carbon uptake, probably in the high latitudes, also known as the missing sink (Chapter 8). The natural uptake of carbon might be a result of the longer growing season, or of direct fertilization of the plants by rising atmospheric CO_2 concentrations. Or it could be fire suppression, or fertilization by the nitric acid component of acid rain. It's not known.

The land biosphere might reverse itself and release carbon in the future as a result of increasing respiration rates in soils. Soil organic carbon comes from plant material, mostly: leaves and roots. How much of it accumulates in the soil depends on how quickly the plants are growing, but also on how quickly the organic material degrades. Soil organic carbon is degraded by soil bacteria and fungi, which work more quickly when the temperature is warm. Soils in the tropics have less organic carbon in them than soils in the higher latitudes (Figure 16), because the carbon decomposes so quickly in the tropics. Think about a ham sandwich sitting on the ground for a few days, in the tropics and in the tundra. Which sounds tastier?

Permafrost soils figure prominently in news stories and books about climate change, because melting permafrosts have had a visible impact on the landscape already, in the Arctic in particular. Permafrost soils often contain high concentrations of organic carbon, which have been protected from degradation by their frozen condition. In particular, there are deposits of nearly pure organic matter called peats. Coal is derived from ancient fossil peats that have been buried and cooked in the Earth's interior.

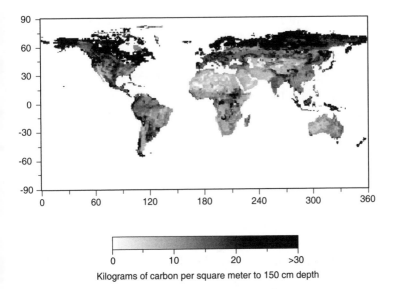

Figure 16. Map of soil organic carbon concentration. Colder soils store more carbon, as indicated by the darker shading. From ISLSCP Initiative II, NASA.

The peats in Arctic permafrost soils begin to decompose when they thaw. Even if they have been frozen for thousands of years, they contain viable microbial spores that blossom into a bacterial community as soon as conditions are right. Decomposing peats give off a mixture of the greenhouse gases CO_2 and methane.

When the possible effect of temperature on soil carbon respiration is included in a climate model, the effect is to amplify the original global warming forcing. This is not a well-constrained estimate, because it is difficult to predict how strong the response will be. But model projections are that the feedback could be significant.

The terrestrial biosphere might respond to climate on timescales of decades or a century. Ultimately, the amount of amplification that the terrestrial landscape can offer is limited by the amount of soil carbon available, about 2000 Gton C. There is

5000 Gton C of coal, and even if this were all burned, the most that the soil could offer would be perhaps 1000 Gton C (half of the available) or less. So the terrestrial biosphere has the capacity to be a significant carbon amplifier on the short term, but on the longer term the amount of carbon it has to offer is limited.

The conclusion from analysis of the ice core data in Figure 15 was that CO_2 feedbacks might increase the original warming by 15–80% (Scheffer paper in Further Reading).

The oceans today are taking fossil fuel CO_2 out of the atmosphere, as discussed in Chapter 8. On longer timescales, the oceans may begin to contribute carbon to the atmosphere, in response to climate warming.

The amount of CO_2 that dissolves in seawater depends on temperature. Like other gases, CO_2 tends to degas as the water warms. A glass of cool tap water, left to sit and warm up in the Sun, will form bubbles on the sides of the glass as the dissolved gas is forced out of solution. Simple calculations, and more complex model simulations of the ocean carbon cycle, predict an atmospheric CO_2 increase of about 10–15 ppm per °C of warming in the ocean.

The increasing atmospheric CO_2 at the end of the ice age (Figure 8) probably originated in the ocean, although no one is really sure how the ocean accomplished this trick. The change in CO_2 is far greater than can be explained by the temperature effect on CO_2 solubility alone. Some of the CO_2 rise may have been caused by shifting ocean circulation. Scientists have a reasonably good idea how the circulation of the ocean in glacial time differed from its circulation today. The ocean circulation could change in the future as well, but this is not so easy to forecast.

Further back in time, in the middle of the glacial climate about 50 to 30 thousand years ago, proxy tracers of the temperature of the deep sea showed periods of relative warmth lasting about 5 thousand years each. These warmings correlate with slightly elevated CO_2 concentrations in the atmosphere (Figure 8). The CO_2 rise through these events is consistent with the warming effect on CO_2 solubility.

The oceans are probably a slower carbon feedback than the terrestrial biosphere, because it takes a long time to warm up the deep ocean. The ocean's contribution to the carbon amplifier may take millennia to play out completely. Simple calculations of the potential impact of ocean warming on the carbon budget do not look too terribly apocalyptic, but until the CO_2 rise at the end of the last ice age is understood, it will be difficult to be too confident about a forecast for the future.

The largest potential CO_2 amplifier is also probably (hopefully) the slowest. There is an enormous amount of carbon on Earth in the chemical form of methane, frozen into a type of water-ice called clathrate or hydrate deposits. Hydrate can form from water and almost any gas. There are CO_2 hydrates on Mars, while on Earth most of the hydrates are filled with methane. Most of these are in sediments of the ocean, but some are associated with permafrost soils.

Methane is thirty times more powerful than CO_2 as a greenhouse gas, molecule for molecule. Released to the atmosphere, it degrades to CO_2, another greenhouse gas, in about a decade. As CO_2, it accumulates in the atmosphere, the same as fossil fuel CO_2 does.

Methane hydrate deposits would seem intuitively to be the most precarious of things. Methane hydrate melts if it gets too warm, and releases methane at atmospheric pressure even if the temperature is cold enough to maintain the water as ice. Hydrate in ocean sediments would float to the ocean surface and melt if it were not buried in mud.

And there is a lot of it, thousands of Gton C of methane, as much as all the rest of the traditional fossil fuel deposits. If just 10% of the methane in the hydrates were to reach the atmosphere within a few years, it would be the equivalent of increasing the CO_2 concentration of the atmosphere by a factor of 10, an unimaginable climate shock. The methane hydrate reservoir has the potential to warm Earth's climate to Eocene hothouse conditions, within just a few years. The potential for planetary devastation posed by the methane hydrate reservoir therefore seems comparable to the destructive potential from nuclear winter or from a comet or asteroid impact.

Most of the methane hydrate on Earth is in sediments at the bottom of the ocean. Of that, most of it is dispersed, spread out in small concentrations over wide areas of the sea floor. Most of the methane is derived from fermentation of organic carbon from plankton buried millions of years ago, now hundreds of meters below the sea floor.

The melting point of hydrate is similar to that for regular water ice, and a temperature change of a few degrees in the sediment column would melt a lot of it. However, it will take a long time to warm up the places where the hydrates are. Hydrate is mostly hundreds of meters deep in the water column of the ocean, which will take centuries to warm, and under a few hundred meters of mud, which will slow things down another thousand years or so. The Arctic is a special case, because the water is cold enough for hydrates to exist as shallow as about 200 meters water depth. The Arctic is also warming more intensely than the global average because of the feedback from melting sea ice. Even so, the timescale for melting the Arctic hydrates is decades to centuries.

The question is whether the released methane can escape to the ocean or even to the atmosphere, or will it remain trapped in the sediment column. If hydrates deep in the sediment column melt, the methane released might rise into cooler temperatures shallower in the sediment column. The surface sediment might act like a cold trap to prevent the released methane gas from escaping. However, there is evidence that the methane can manage to get through the cold trap and into the overlying ocean. Seismic studies show "wipeout zones" where all of the layered structure of the sediment column through the stability zone is gone. These are thought to be places where explosions of escaping gas have broken through the sediment column. The sediment surface of the world's ocean has thousands of craters in it called pockmarks, interpreted to be what these gas explosions look like from the surface. It could be that the methane burps are too fast for the gas to freeze into hydrate, or it could be that fluid flowing upward with the gas carries with it heat, preventing the methane from freezing.

And there is the possibility of landslides. When hydrate melts and produces bubbles, there is an increase in volume. The newly formed bubbles might lift the sediment grains off of each other, so that they don't stick together as well, allowing the sediment column to slide. The largest submarine landslide known is off the coast of Norway, called Storegga. The Storegga slide itself occurred about 2–3 millennia after a warming in the water at the end of the last glacial time. The slide excavated on average the top 250 meters of sediment over a swath hundreds of kilometers wide, stretching half-way from Norway to Greenland. There have been comparable slides on the Norwegian margin every approximately 100 kyr, reminiscent of the glacial cycles, although the dating is not very precise. It is not certain that melting hydrate was the culprit; the slide could also have been caused by glacial debris piling up on the sea floor after the melting of the European ice sheet.

Even if methane escapes the sediment column, it is not so easy for it to make it to the atmosphere without being degraded to CO_2. Methane can leave the sediment in three possible forms: dissolved CH_4, bubbles of gas, and frozen hydrate. Dissolved methane is chemically unstable in the oxic water column of the ocean, but it might persist for a few decades, long enough perhaps to evaporate from the surface ocean to the atmosphere.

Bubbles of methane are typically able to rise only a few hundred meters before they dissolve. A landslide may also release solid hydrate, the icy form, which floats in water just like regular ice would. Floating hydrate could carry methane to the atmosphere more efficiently than bubbles can.

The Storegga landslide probably did not release a catastrophic amount of methane, based on the volume of the slide and the amount of hydrate one might reasonably expect in it. Even if all the methane from Storegga managed to reach the atmosphere, it would have had a smaller climate impact than a volcanic eruption. Records of atmospheric methane concentration from ice cores don't show any spike in methane concentration at this time. So far, no one has proposed any scenario by which a large fraction of the methane in ocean hydrates could escape to the atmosphere all at once.

Methane hydrates are sometimes associated with permafrost deposits, but never too close to the soil surface, because hydrate requires methane at high pressure. Sometimes freezing, flowing groundwater creates a sealed layer of ice in the soil, which increases the pressure in the pore space below and traps methane. The great depth and the ice seal help insulate the hydrate from climate changes at the soil surface.

The most important means of eroding such a sealed ice layer is sideways, from the coast, where the subsurface ice is exposed to the waters of the ocean. As the ice melts, the land collapses, exposing more ice in a melt-erosion process called thermokarst erosion. The northern coast of Siberia has been eroding in this way for thousands of years, but rates are accelerating. Entire islands have disappeared in historical time. There are high concentrations of dissolved methane in the waters of the Siberian shelf, indicating escape of methane from coastal erosion into the atmosphere, probably from the degradation of peat. Total amounts of clathrate methane in permafrost soils are very poorly known, with estimates ranging from 10 to 400 Gton C.

Methane hydrates are blamed by some for the Paleocene Eocene thermal maximum event, which happened 55 million years ago and was discussed way back in Chapter 6. Hydrates are an attractive hypothesis because methane is isotopically very "light" (lots of carbon-12, not much carbon-13). It doesn't take a ridiculous amount of methane to explain the light carbon spike of the PETM. If the carbon source were isotopically more similar to the atmosphere or ocean carbon, it would take more total molecules of CO_2 to explain the isotopic change. I personally don't believe that methane was the source of the PETM carbon, because lots of carbon is required to explain how much warmer the ocean got, and how much $CaCO_3$ dissolved in the acid of the CO_2. But it's fair to say that the jury is still out on this question.

Some climate scientists speculate about the role of methane hydrates in the glacial cycles. The deep ocean warmed and cooled several °C through the cycles, colder during glacial time. Measurements of the methane concentration in the atmosphere, from ice core bubbles, show some variations but not enough to drive the large temperature changes through the glacial cycles. It was CO_2 driving the greenhouse effect, mostly. There is also the constraint that methane has a peculiar light isotopic ratio of carbon-12 to carbon-13, so any large methane release would leave traces in the isotopic composition of the $CaCO_3$ forming in the ocean. No really large methane releases are seen.

We can hope that the lack of a methane response to the end of the last ice age means that future warming will also have no impact on the methane. But there is an important difference between the past and the future. The future ocean could be warmer than it has been in millions of years. It takes millions of years for methane to build up into hydrates, because methane is produced only very slowly in the sediments. The warming into our current interglacial climate released only a small amount of methane, maybe because the ocean was already this warm just a hundred thousand years ago, and not much extra hydrate accumulated in the cold times since then. If the ocean gets warmer than it has been in millions of years, hydrate may melt in places where it has been building up for a long time.

The hothouse climate of 40 million years ago probably did not have much methane hydrate in it. The hydrates we have today probably grew in the cooler times more recently than that. If the Earth returns to a hothouse climate, it seems inevitable that it would eventually lose most of that methane. It is an open question how long this would take, and how much of the methane actually manages to re-join the carbon cycle to affect climate. Hydrate stability calculations suggest that hydrates could ultimately release as much carbon as the CO_2 released from fossil fuels, doubling the long-term climate impact of global warming.

Climate records of the past give reason to fear that the carbon cycle could eventually act as an amplifier of human-induced climate change. The glacial cycles, for example, were apparently instigated by wobbles in the Earth's orbit, with the carbon cycle playing some kind of amplifying role. The biosphere responded to warming by exhaling CO_2, and to cooling by inhaling.

Global warming differs from climate changes through the glacial cycles in that now the initial cause for the climate change arises from CO_2, rather than a direct thermal driver such as the orbit wobbling. The carbon cycle is responding to all the extra CO_2 by inhaling, into the ocean and into the high-latitude land surface.

On timescales of centuries and longer, the lesson from the past is that this situation could reverse itself, and the warming planet could cause the natural carbon cycle to exhale CO_2, amplifying the human-induced climate changes, rather than damping it down as it is doing at present.

CHAPTER 11

Sea Level
in the Deep Future

T he most compelling reason to worry about sea level rise in the future comes from sea level estimates from the past. Ancient coral reefs and relic beach deposits attest to large changes in the sea level associated with past climate changes. Periods of low sea level, such as the last glacial maximum, are more difficult to document in this way because the relic seashore is now submerged under more than a hundred meters of water. In some locations such as Barbados, the land surface itself is rising out of the ocean faster than the flooding of the ocean, leaving the traces of the ancient shorelines exposed for scientific inspection. As described in Chapter 4, oxygen isotopes in $CaCO_3$ shells deposited in the deep ocean also carry information about ancient sea levels.

Sea level during the last glacial time, 20 millennia ago, was about 120 meters lower than today (Chapter 5). Most of that missing water was tied up in the North American and European ice sheets. The temperature at this time was about 5–6°C colder than today, on a global average. Figure 17 shows the difference in temperature on the horizontal axis, and the change in sea level on the vertical axis, for this and other climate changes past and future.

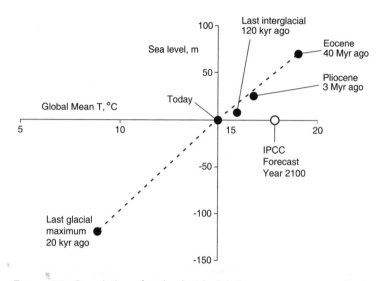

FIGURE 17. Covariation of sea level with global average temperature in the geologic past, compared with the IPCC forecast for sea level rise by the year 2100.

During the last interglacial climate, about 120 millennia ago, sea level was 4–6 meters higher than today. The atmospheric CO_2 concentration was comparable to the pre-anthropogenic value of about 280 ppm. The ice melted primarily because the Earth's orbit brought warmer summer temperatures to the northern hemisphere. Overall, the temperature was at most 1°C higher than the recent, preanthropogenic climate, while temperatures in the high northern latitudes were perhaps 3°C warmer than preanthropogenic. The global temperature at this time was comparable to that of global warming so far, but falls far short of projected climate changes in the coming centuries.

Before the start of the glacial / interglacial cycles, about 3 million years ago, was a time period called the Pliocene. The Antarctic ice sheet existed at this time but it was smaller than today, and the Northern hemisphere was ice-free, so sea level was 20–25 meters higher than it is now. The global average temperature is documented most clearly in oxygen isotope and trace element concentrations in deep ocean $CaCO_3$. These measurements indicate that the Earth was perhaps 2°C warmer at that time.

The Antarctic ice sheet is about 15 million years old. For a long time before its formation, the Earth was in an ice-free "hothouse" climate state (Chapter 6). The height of the hothouse climate came in the Eocene time period, about 40 million years ago. The concept of sea level begins to lose its definition on timescales this long, because the continents themselves are floating in the viscous fluid of the Earth's mantle. The continents viewed in an ultra time-lapse movie spanning millions of years would appear to be bouncing up and down like rafts in choppy waters. The way to deal with this difficulty is to base the sea level change for the hothouse climate of the Eocene in Figure 17 on the sea level change that would result from melting all the ice on Earth today, in the continents' current configuration. If we were to go ice-free today it would result in about 70 meters of sea level rise. The temperature during the Eocene was 4–5°C warmer than prehistoric, again based on analyses of the chemistry of deep ocean $CaCO_3$ deposits.

There is a world of complexity and richness summarized in the data points from the past in Figure 17. The accumulation of water into ice sheets is governed by the circulation of the atmosphere, which is itself driven by the ocean currents and the distribution and elevation of the continents. The physics of ice sheet flow, determining how large the sheet can grow, is only rudimentarily understood. The responses of the various ice sheets to changes in climate have been the focus of life's work and ambitions for countless hardworking scientists. Without meaning to denigrate all of that hard work and natural wonder, there is also a virtue in simply connecting the dots, as has been done in Figure 17. Sea level in the past varied clearly and strongly with changes in global average temperature.

Also plotted on Figure 17, and in contrast to the data points from the past, is the forecast sea level rise for the year 2100. The temperature change in the coming century could be 3°C, which is high enough, according to the Earth system's behavior in the geologic past, to raise sea level by perhaps 50 meters. But the IPCC forecast of sea level change in the coming century is only about half a meter. A temperature change this large in the past looks like it would have been accompanied by a sea level change one hundred times higher than that.

The difference between the past and the forecast for the coming century is the assumption that it will take longer than a century to really change the sea level by very much. Sea level rises in a greenhouse world for reasons that tend to take millennia to come to fruition. Given the long atmospheric lifetime of fossil fuel CO_2 described in Chapters 8 and 9, however, it is clear that there is lots of time available to melt ice. The sea level rise one thousand years from now will be much higher than what the coming century will bring. The forecast to 2100 is only the tip of the iceberg. So to speak.

Part of the sea level rise is caused by thermal expansion of the ocean. Warming seawater expands and takes up more room, pushing sea level up like mercury in a thermometer. Thermal expansion is probably the easiest component of sea level rise to forecast, and for the coming century it is expected to be the largest component, contributing about a quarter of a meter. It takes perhaps a millennium, though, certainly longer than a century, to fully change the temperature of the deep ocean. The potential millennial-timescale average temperature change from global warming is about 3°C (Chapter 8). This translates to about 1.5 meters of sea level rise, ultimately, from thermal expansion of seawater.

On longer timescales of millennia, the dominant agent of sea level change is melting land ice. Only land ice can change sea level. Floating ice already displaces its own weight in seawater, as realized by Archimedes as he soaked in his bathtub. When floating ice melts, it just exactly fills in the hole it made in the water. Based on Figure 17, melting land ice could eventually increase the sea level by perhaps 50 meters. Most of this water comes from large ice sheets in Greenland and Antarctica (Chapter 3).

The temperatures over Antarctica are comfortably below freezing, now and for the foreseeable future. Greenland however is close to the melting point. Models, corroborated by data from the last interglacial period 120 millennia ago, predict that the Greenland ice sheet would melt significantly if the summertime temperatures were 3°C warmer than they are today. According to the models, it ought to take several millennia to melt the Greenland ice sheet. However, there are reasons to worry that real ice sheets can melt in ways that would evade the current state-of-the-art model forecasts. As discussed in Chapter 3, ice knows a few tricks for melting quickly that glaciologists are not predicting in advance, but only discovering as they happen.

No one predicted the explosion of the Larsen B ice shelf. Sometime within a few days following March 5, 2002, a 200-meter-thick floating shelf of ice the size of Rhode Island exploded into icebergs. It is thought that the integrity of the ice shelf was undermined by meltwater digging vertical cracks called crevasses into the ice. If the crevasses carved the ice into blocks that were taller than they were wide, they would tend to tip over like a line of dominos, enabling the entire shelf to blow out into icebergs within a short time.

The ice in an ice shelf was already floating in the ocean, so the fragmentation and subsequent melting of all that ice has no direct impact on sea level. However, it appears that ice shelves act to impede the flow of ice from the interior ice sheet into the ocean. The Larsen B ice shelf was fed by ice streams, rivers of rapid ice flow within the ice sheet. The velocity of these ice streams, measured by GPS, has accelerated by as much as a factor of seven following the explosion of the ice shelf. In Greenland, the flow of ice streams in the Jacobshavn glacier also accelerated, by a factor of two, after the ice shelf at the outflow of the glacier broke up in 2002. The West Antarctic ice sheet, described further below, flows into the Ross Ice Shelf. Meltwater ponds have started forming on the Ross Ice Shelf, foreshadowing a possible catastrophic breakup such as occurred to Larsen B. If the West Antarctic ice sheet collapsed completely, it would raise sea level by about seven meters.

No one can explain the Heinrich events that punctuated the climate during glacial time 30 to 70 thousand years ago. First discovered as layers of ice-rafted sand in sediments of the Atlantic Ocean, the Heinrich events document the collapse of the Laurentide Ice Sheet in North America into armadas of icebergs, raising sea level by several meters within a few centuries. Icebergs are an extremely efficient way of melting an ice sheet, because they transport the ice from the high latitudes, where sunlight is weak, down to low latitudes where the sunlight is more intense.

The Laurentide Ice Sheet flowed into the ocean at a latitude of about 60° North. The Greenland Ice Sheet today is centered at about 70° N. Greenland is a warm ice sheet, close to the melting temperature, and getting warmer. Because the mechanisms by which the Laurentide dumped so much ice into the ocean are not understood, it is impossible to predict whether the Greenland Ice Sheet could pull the same Heinrich maneuver. If such a collapse should start, it would be impossible to stop.

The fate of an ice sheet is determined by its temperature at the bed, in particular whether the ice near the bed has reached the melting temperature. When the ice is frozen securely to the bed, the ice can flow only by deforming, the surface flowing while the bed stays put. When the base melts, it lubricates the bed and allows the entire column of ice to flow without deforming. Models tell us that once an ice sheet starts melting down, the friction from the flow generates heat, which facilitates faster flow. A little water at the bed is all it takes for an ice sheet to lose its footing and plunge into the sea.

In the current models of ice sheet flow and accumulation, the main way for heat to travel from the atmosphere to the base is by thermal conduction through hundreds of meters of ice. It would take millennia for the flow rate of an ice sheet that worked like this to respond to a change in climate. The real Greenland ice sheet, in contrast, responds within just a few months to changes in surface conditions. The flow accelerates in the summer and lags in winter, with no 1000-year lag.

The surface apparently sends its heat down to the base by means of flowing melt water. Glaciologists in the field can watch water submerging into holes in the ice called moulins. The mystery is how the water is able to penetrate through the ice to the bed without refreezing. Most of the ice column is below the freezing point. Because the mechanism for getting the water unfrozen through the ice is not understood, models are not yet able to reproduce this trick.

The other large, potentially unstable mass of landed ice is called the West Antarctic Ice Sheet, large enough to raise sea level by about 5 meters were it to melt. The good news is that unlike Greenland, the surface air is below freezing all the way down to sea level in this region, so an increase in air temperature is less likely to generate melt water.

The West Antarctic Ice Sheet flows to the ocean through a parallel series of ice streams. Ice flows like water, very viscous water flowing extremely slowly. Typical flow rates of a stable, well-grounded ice sheet or glacier may be a meter per year, while an ice stream flows a hundred times faster than this. It is thought that the ice in a stream flows so easily because it is floating on a layer of meltwater, or perhaps a mud composed of liquid water and sand or small rocks.

There are five ice streams draining the West Antarctic Ice Sheet into the Ross Sea, but only four of them are active today. One of them, ice stream C, suddenly stopped flowing quickly, about 150 years ago; glaciologists know about it only by seeing its relic structure in the ice. Ice streams are known to start and stop their mad dash to the sea without much warning on their part or understanding on ours. The West Antarctic ice streams flow into the Ross Ice Shelf, which is beginning to produce the same meltwater ponds that were seen on the Larsen B ice shelf before it exploded in Figure 6.

The West Antarctic Ice Sheet is the world's only entirely marine ice sheet. Nowhere does it touch the ground above sea level. A curious reader may wonder how such a thing came into being. Could ice have formed at the bottom of the ocean, or did it grow on the ocean surface, eventually to reach the sea floor?

The answer is that when the ice sheet first started forming, the rocks were above the ocean surface. The weight of the accumulating ice sheet caused the land surface to sink in a process called isostatic depression. Just like sea ice or a boat in the water, the crust of the Earth floats in the extremely thick fluid of the mantle rocks below. When weight is added to the crust, it sinks down a bit, like when a person steps onto a dock. Because the rocks flow sluggishly, isostatic elevation changes take tens of millennia. This topic has come up before, in our previous discussion of sea level rise in Chapter 7.

The crust sank most deeply in the center of the ice sheet, just as a floating boat dock would sink down most strongly under where the person stepped on it, if it were somewhat flexible. The result is that the ice sheet is anchored in increasingly deep water as you go into the center. Instead of sitting atop a hill as we might have expected, the ice sheet is filling a hole (Figure 18).

An ice sheet in this configuration may be susceptible to a runaway collapse into the ocean. The ice flow today is impeded by the presence of the Ross Ice Shelf, and by a few places where the

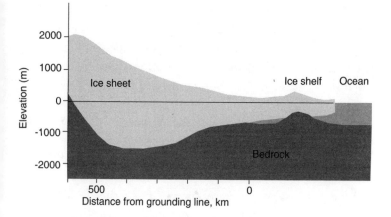

FIGURE 18. Cross section of the West Antarctic Ice Sheet, from Oppenheimer, *Nature* 393, 325–333, 1998.

bedrock rises up to impinge on the ice shelf. If the Ross Ice Shelf collapses, the resulting acceleration of ice flow to the sea could lead to thinning of the ice sheet. As the ice grew thinner, it would tend to float higher in the ocean like a floating dock when a person steps off. The base of the ice sheet could unmoor from the bed, allowing the ice to flow even faster. The land surface will tend to rise also, for the same reason as the thinning ice did (and the floating dock), but this process takes thousands of years, much slower than the ice rebound, because rocks flow more slowly than water does.

The bottom line is that like Greenland, the West Antarctic Ice Sheet seems capable of switching to a faster mode of melting that would be difficult to forecast in advance, and impossible to stop once it got underway.

There are two reasons why we are interested in how quickly the Earth's ice can melt. First, if an ice sheet can collapse in a century in the past (as it did during Heinrich events or Meltwater Pulse 1A), it could do so again in our century, affecting the amount of sea level rise that we and our descendants would actually live to see. Some of the climate changes in the past happened very quickly, but atmospheric CO_2 concentration has never gone up as quickly as it is now. The forecast for the coming century is based on models. If the models are too slow, the forecast is too low.

The other reason to care about ice melting rates is that if an ice sheet can collapse in a millennium instead of 10 millennia, then it will be quick enough to respond to the much warmer first millennium of global warming. The first thousand years is the hottest because the CO_2 is slowly being absorbed back into the Earth by weathering reactions. If ice sheets take many thousand years to melt, then they may survive that hot first millennium.

It appears from Figure 17 that a 3°C difference in global mean temperature has a huge impact on sea level, somewhere in the neighborhood of 50 meters. If the ice melts in a millennium, then a fossil fuel release of 2000 Gton C would elevate temperatures to above 3°C for that long. If on the other hand the ice takes 10 millennia to melt, then 5000 Gton C would be required to ultimately melt that much ice

Combine the amount of fossil fuel carbon available, the long lifetime of released CO_2 in the atmosphere, and the sensitivity of ice sheets to global climate on long timescales, and you conclude that humankind has the potential to increase the sea level by tens of meters. Even under the best of assumptions, it would appear that an ultimate sea level rise of at least 10 meters might be difficult to avoid, unless CO_2 is actively removed from the atmosphere.

As discussed in the epilogue, a 2°C warming of the global average is often considered to be a sort of danger limit benchmark. Two degrees C was chosen as a value to at least talk about, because it would be warmer than the Earth has been in millions of years. Because of the long lifetime of CO_2 in the atmosphere, 2°C of warming at the atmospheric CO_2 peak would settle down to a bit less than 1°C, and remain so for thousands of years. The last interglacial period, 120 thousand years ago, was about 1°C warmer than the presumed natural climate of 1950. Sea level was 4 or 5 meters higher at that time. The larger trend, extending from the glacial low to the Eocene high sea level states, is 10 or 20 meters of sea level change per °C temperature change. If global warming buys into that trend, sea level could rise by 10 meters or more, even if the maximum global warming is limited to only 2°C.

So what's this all going to cost? It's only real estate, after all. The economic value of real estate, or anything else for that matter, is not an intrinsic property of the thing, like density or temperature, but rather its value is determined by what people are able and willing to pay.

To think about the economic value of land, I find it helpful to think in terms of years of economic production. It takes a few decades for a person to pay off a mortgage, in general. The value of real estate is limited by the constraint that it must be possible to pay for a place to live in less than a human lifetime. If it took a millennium to pay off a house, no one could afford it because no one lives that long. Let's say that the economic value of the land on Earth is equal to 10 years of economic production.

The 2.2% of the land surface that would be flooded by 10 meters of sea level rise is currently home to about 10% of the world's population. Cities are disproportionately located in the lowest parts of the land. Low-lying river deltas are agriculturally fertile. These properties must have higher than average value to the human race. Let's scale the price of land according to its population and say that 2.2% of the land surface is worth 10% of our real estate holdings, because it houses 10% of the world population.

Economically, it would seem that this loss would be equal to about one year's worth of economic production (the whole surface = 10 years of production, times 10% of the area is flooded). If the sea level rise is spread out over a century, then the economic hit is only 1% of global GDP per year, not too expensive at all.

The flaw in this reasoning is that we cannot really buy back the land surface for its economic trading value. Some of the land surface can be defended against the encroachment of the ocean, like the Netherlands. But the defense cost of land is not the same as its trading cost. Much of the land could not be defended at any cost, like coralline islands where the ocean seeps up into the groundwater.

If we do the math, 50 meters of sea level rise resulting from say 5000 Gton of carbon works out to about 10 cm^2 of area lost per kilogram of carbon. A gallon of gasoline burned wipes out maybe 50 cm^2. Every year, the average American emits enough CO_2 to ultimately flood 100 square meters, almost 1000 square feet, the size of a luxury Paris apartment.

Ultimately, the item under negotiation here is the long-term carrying capacity of the planet. The selling price is short-term convenient energy. The clearest long-term impact of fossil-fuel CO_2 release may be the sea level rise associated with melting of the great ice sheets. We have the capacity to ultimately sacrifice the land under our feet.

CHAPTER 12

Orbits, CO$_2$, and the Next Ice Age

Two centuries ago, climatologists were more concerned about the next ice age than they were about global warming. Svante Arrhenius, who first estimated the climate sensitivity to atmospheric CO$_2$ in 1896, was interested in explaining the cause of the last ice age. From moraines it was known that the landscape had undergone repeated assault from massive ice sheets. The timing was not known very well, because moraines are difficult to date, especially without carbon-14 dating, and they document only the coldest times, while the warmest times that we're interested in, the interglacials, leave no trace. So the landscape offered no information about the longevity of warm climate intervals such as ours, only an ominous drumbeat of glaciations in the past.

The 1970s saw the development of time records from the ratio of oxygen-18 to oxygen-16 in ocean sediments (Chapter 5). It was found that the Earth spent most of the past million years or so in a glacial climate state. Interglacial climate stages, the most recent ones anyway, generally lasted for 10 millennia, while glacial states might last ten times as long. The orbital theory of climate explained that the duration of a warm interval was determined by the orbital cycle called precession of the equinoxes, which has a half-cycle time of about 10 millennia. And guess what. The current interglacial period has lasted about 10 millennia. Are we near the end of the line?

The nucleation of an ice sheet is the part of the glacial / interglacial climate cycle that scientists understand relatively well, because it starts from a climate like our own. We know the humidity and the cloud cover and most everything else about the interglacial climate state. The previous climate state in the glacial cycle was the last glacial climate twenty thousand years ago, which has to be reconstructed using pollen grains and oxygen isotopes and other proxy methods.

During the glacial inception from the last interglacial period 120 thousand years ago, CO_2 remained high, at a typical interglacial level of 280 ppm, until after the ice sheets started to grow. If it wasn't a drop in CO_2 that caused the ice sheet to form, the other usual suspect would have to be a change in Earth's orbit. The intensity of sunlight varies in different rhythms at different latitudes and around the seasons, but the amount of ice on Earth seems to listen particularly well to the intensity of sunlight in the northern hemisphere summer, at about 65° N latitude. Figure 19 shows that whenever the northern hemisphere summer sunshine gets dimmer than a particular trigger value, the amount of ice on Earth grows and sea level falls, without exception through the last 800 thousand years.

It makes sense that the summer would be the critical time for the development of an ice sheet, because the winters are always cold enough to snow up in northeastern Canada. The fate of an ice sheet is determined by whether the summers are warm enough to melt the snow or not. Summertime sunshine may not be the only factor that determines the nucleation of an ice sheet, but it sure seems like the most important factor. The northern hemisphere summer is the solar-forcing sweet spot that

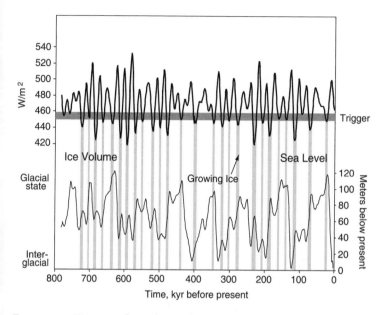

FIGURE 19. Top: Northern hemisphere summer sunshine intensity as modulated by orbital variation. Bottom: ice volume. Vertical bars are times when summer sunlight drops below a Trigger value. In those times, ice grows.

drops the entire planet into an ice age, like a sucker punch to the solar plexus.

As a matter of fact, summertime sunlight at 65° N is getting a little thin lately. It's approaching the trigger value now (Figure 20), and will almost graze it about three thousand years from now. For what it's worth, the trigger model decides not to glaciate this time around, and instead continues in an interglacial mode for another fifty thousand years. But the real world is fuzzier than the trigger model. Surely in the real world it must matter, at least a little bit, how long the period of cold summers lasts, and how much it snows in the winter. The onset of an ice age is mostly triggered by summer sunlight intensity, but in close calls like this one the secondary causes might also be important. Also, even if the trigger mechanism were all-powerful, the actual trig-

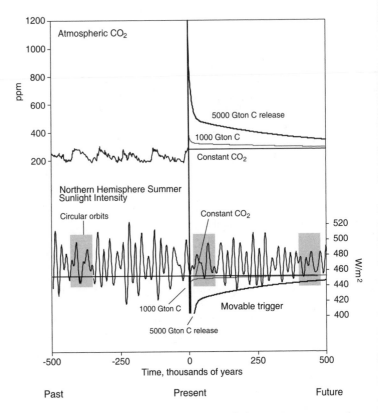

FIGURE 20. Top: atmospheric CO₂ past and future. Bottom: northern hemisphere summer sunlight intensity, with the value of the trigger as it is altered by atmospheric CO₂ concentration.

ger value for the real world can be diagnosed from the ice volume record, but not without a lot of uncertainty. In the model, if we assume a slightly higher trigger value, the near miss could easily be changed into a hit, the initiation of the next ice age.

This is a case where a reliable forecast cannot be made with the tools at hand. It's too close to call. There is nothing in the orbital wobbles that suggests that an ice age is going to start next week. It could be that a new ice sheet would naturally start to form sometime in the next millennium or two, or it could be that nothing would have happened, even with global warming, for tens of millennia. According to the trigger model, if the climate system misses the glacial express this time, the next opportunity will be 50 thousand years from now.

The next hundred thousand years will be a time when there is less variability in solar forcing than usual. This is because the Earth's orbit is nearly circular today (Figure 7). Much of the change in sunlight intensity that drives climate arises from an interplay between the tilt of the Earth and the elliptical orbit. The northern hemisphere summer comes when the North Pole leans toward the Sun. Around the elliptical orbit, northern summer can happen when the Earth is close to or far away from the Sun. If the Earth is far away, it will be a cool summer.

Times like today, when the orbit is nearly circular, diminish the solar variability because it doesn't matter where in the orbit the Northern hemisphere summer comes, it's about the same distance to the Sun all around the orbit. The last time the Earth was in this configuration was about 400 millennia ago. The interglacial period at that time was about 50 thousand years long, the same as what the trigger model is predicting for our current interglacial period.

Bill Ruddiman argues in a delightful book entitled *Plows, Plagues, and Petroleum: How Humans Took Control of Climate* that human activity began changing the atmospheric CO_2 and methane concentrations thousands of years ago by clearing land for agriculture. If CO_2 had followed its natural trajectory, he says, the next ice age would already have started. His conclusion is based on CO_2 concentration changes 120 thousand years ago, through the last interglacial time. The highest CO_2 concentration came at the beginning of the interglacial time period, and CO_2 decreased through the interglacial period until it was time to glaciate again. CO_2 in our interglacial started out high, then dropped, heading upward again about eight thousand years ago. That last uptick Ruddiman blames on early human agriculture.

Since the publication of Ruddiman's book, ice core records of climate have been extended farther into the past, reaching now the time interval 400 thousand years ago when Earth's orbit around the Sun was nearly circular as it is now. The interglacial period at that time lasted 50 thousand years, and atmospheric CO_2 remained high throughout that time. This would seem to imply that there is not necessarily a 10 thousand year time limit for interglacial periods, so our current interglacial was not necessarily doomed as Ruddiman imagined. Also, the ratio of carbon-13 to carbon-12 in atmospheric CO_2 through the rise of the last few thousand years seems to indicate that the CO_2 did not come from cutting trees, but rather from the oceans. At any event, both of us agree that humankind has the potential to take control of the ice ages in the future.

All of the glacial inceptions from the past took place with atmospheric CO_2 concentration at typical interglacial values of 260–280 ppm. CO_2 today has risen to 380 ppm, and continues to go up. How will this affect the onset of glaciation? My collaborator Andrey Ganopolski and I used a model of the climate called CLIMBER to answer this question. The model has an atmosphere, an ocean, and ice sheets that form, flow, and melt.

Andrey set the model up with an initial climate much like today's, with the northern hemisphere summer coming when Earth is farthest from the Sun. He then slowly dialed up the eccentricity of the orbit, decreasing slowly the intensity of sweet-spot sunshine. When the model atmospheric CO_2 is 280 ppm, the model begins to grow a Canadian ice sheet when summer solar flux drops below 455 W/m^2. This is very close to the trigger value derived from the paleoclimate record. When Andrey increases the atmospheric CO_2 concentration to, say, 560 ppm (doubled the natural value) and runs the experiment again, the Earth is able to stay out of a glaciation until summer sunlight drops to only 407 W/m^2 (Figure 21).

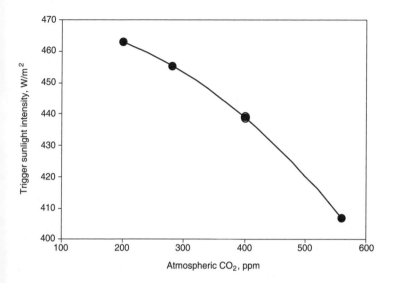

FIGURE 21. The trigger northern hemisphere sunlight intensity as a function of atmospheric CO_2 concentration. From Archer and Ganopolski, *Geochem., Geophys., Geosys.* 6(5): doi 10.1029/2004GC000891, 2005.

Different climate forcings can be compared according to their intensity in watts per square meter of surface, W/m^2 (Chapter 2). The climate / ice sheet model predicts that for the job of nucleating an ice sheet, 1 W/m^2 of warming from CO_2 is able to compensate for about 4 W/m^2 of solar cooling in the northern hemisphere summer. It makes sense that orbital solar forcing would be weaker at starting an ice age than CO_2 forcing, because the orbit doesn't change the total amount of sunlight the Earth receives, averaged over the year, by very much. Orbital variations mostly change the distribution of sunlight, among different latitudes and seasons. Dimmer sunshine in the northern hemisphere summer means stronger sunshine in the northern hemisphere winter, for example. The ocean stores enough heat to average out these extremes to some extent. CO_2 forcing of climate, on the other hand, goes in the same direction nearly everywhere, all year round. Therefore 1 W/m^2 of CO_2 forcing should have a stronger effect on the nucleation of an ice sheet than 1 W/m^2 of orbital solar forcing.

The long lifetime of fossil fuel CO_2 in the atmosphere means that human activity will affect the trigger sunshine value for a long time into the future. Atmospheric CO_2 concentrations predicted for the future are plotted on Figure 20, along with the effect of the higher CO_2 concentration on the value of the glacial nucleation trigger. The larger the CO_2 release, the greater the shift in the trigger. Natural evolution of climate, from the natural interglacial CO_2 concentration, was a near miss; it was touch-and-go whether an ice sheet would begin to form or not. With rising CO_2, the trigger moves farther and farther out of reach. It becomes less of a near miss.

So far, the impact we have had on the glacial cycles appears to be relatively small. The northern hemisphere summer sunlight intensity was about 2 W/m^2 brighter than the trigger value with natural CO_2 levels. We have released about 300 Gton C since 1750, which will have the effect of decreasing the trigger value, three thousand years from now, by about another 2.5 W/m^2. The near miss is getting a little less near.

If mankind ultimately burns 2000 Gton C (this is about the business-as-usual forecast for the coming century), then it looks as though climate will avoid glaciation in 50 millennia as well, waiting until the next period of cool summers 130 millennia from now. If the entire coal reserves were used (that is, 5000 Gton C), then glaciation could be delayed for some 500 millennia, half a million years. The Earth could remain in an interglacial state until the end of not our current period of circular orbit, but the next circular time, 400 millennia from now.

On the surface of it, this would seem like a good thing. I live in Chicago, which was covered by a mile of ice during the last glacial time. A return of the ice age would not good for real estate values, I think. I would not put this forecast forward as an argument in favor of CO_2 emissions, however. The potential dangers of warming are immediate, while the potential next ice age in a natural world was not due for thousands of years. If CO_2 release were not already part of our energy infrastructure, no one would argue for deliberate CO_2 release just to stave off the danger of the next ice age. The practical implication of the trigger model is that natural cooling driven by orbital variation is unlikely to save us from global warming.

Another implication, less immediately practical but rather a shift in perspective, is that by releasing CO_2, humankind has the capacity to overpower the climate impact of Earth's orbit, taking the reins of the climate system that has operated on Earth for millions of years.

Carbon Economics and Ethics

Humankind has the potential to alter the climate of the Earth for hundreds of thousands of years into the future. That I feel can be said fairly confidently. But will we? Do social inertia, energy infrastructure, and sheer numbers of people on Earth make the business-as-usual global warming forecast inevitable, or is global warming something that can be avoided? This is much harder to predict. Technologically I believe that it is possible to avoid dangerous climate change, if we so choose. But making a decision: there's the tricky part. Climate change is a global issue that ramps up slowly and lasts for a long time. Negotiating a solution would require a degree of global cooperation that is I think unprecedented in human history.

If the climate of the coming century evolves as it is expected to, or even worse if something unexpected happens, the political will to classify CO_2 as a pollutant will only intensify. Industrial civilization has already shifted its energy source several times in the last few centuries, from wood in the eighteenth century to coal in the nineteenth, then to oil and gas in the twentieth. In the long run, it won't be that hard to change it again. What if human civilization had taken root in a world in which fossil fuels would already have run out by now? Does anyone think that we would be living like the Flintstones, with our feet sticking out the bottom of our automobiles? I have to believe that humankind would have developed workable alternative energy sources if we'd been forced to do it.

Imagine walking down the street in Silicon Valley, California, a $20 bill hanging out of your pocket, trying to find a way to charge your iPod in a world without coal. The engineers are sitting in their cubicles, wailing because they can't think of any conceivable way to charge that iPod and take your $20. This is an unbelievable story, and not just because no one walks in Silicon Valley. Those engineers would come up with something. It is astonishing to sit back and watch market-driven innovation solve problems.

The situation looks a little more daunting, however, when viewed in the global scale over the coming decades. CO_2 emission is closely tied to economic and military supremacy in our world. The Kyoto Protocol, mandating cuts of about 6% below 1990 emission levels, was rejected by the American and Australian governments because it was seen as too invasive to the economy, too expensive to implement. As discussed below, much larger cuts would ultimately be required to stabilize atmospheric CO_2.

The chemistry of the atmosphere is a classic example of a situation called the tragedy of the commons. Individuals profit from releasing CO_2, but everyone collectively pays the price. Each individual's incentive in such a situation is to exploit the common resource to the maximum extent. Cutting CO_2 emissions would require a degree of global cooperation that humankind has never before achieved. In 1985 the Montreal Protocol began the banning of chlorofluorocarbons, or Freons, because they deplete ozone in the stratosphere. This was much

Is this still true?

easier than phasing out CO_2 emission would be, because there were economical alternatives to Freons. There is currently no immediately available alternative to energy from fossil fuels. We would have some work to do first.

How much CO_2 emission is too much? If we were considering building some large building, there might be several potential limiting considerations to how big the building could be. The building should not be so high that it might fall over in heavy wind; it mustn't bring in too much traffic for the streets to bear; perhaps it ruins a view if it exceeds some height, or interferes with air traffic. In the field of engineering, it would be usual to take the time to consider each of these related issues separately. Climate change negotiations are essentially planning exercises in geo-engineering, which can follow the same approach.

The primary climate guardrail is the maximum tolerable change in global mean temperature. If global warming exceeds 2°C, the Earth will be warmer than it has been in several million years. Dialing the climate up to a new one, outside the range of recent natural variability, opens the door to surprises, such as new patterns of atmosphere or ocean circulation, or rainfall and drought, that would be difficult to forecast in advance. Of course an anthropogenic climate change of 1°C would be better than 2°C, and it would probably be safest for humankind to not alter the climate at all. But just as a number to talk about, 2°C is often taken to be a benchmark for a dangerous temperature rise to be avoided if possible in the future.

Sea level rise is another tangible danger limit benchmark. The IPCC forecast for the coming century is about 0.5 meters of sea level rise, which is significant but not generally considered catastrophic. The IPCC business-as-usual sea level rise forecast would seem to leave us on relatively safe ground, at least for the coming century. However, if sea level were to rise by more than a meter or two, it would begin to displace large numbers of people. According to one compilation of population and elevation data (McGranihan et al., http://sedac.ciesin.columbia.edu/gpw/lecz.jsp), a 10-meter rise in sea level would displace about 10% of humanity.

I explained in Chapter 12 why many Earth scientists believe that the IPCC forecast for sea level rise may be an underestimate. The forecast does not include the possibility of the ice sheets responding to warming by increased flow out to the sea. The Greenland ice sheet is beginning to flow faster because of recent warming. Ice shelves have retreated in both Greenland and Antarctica, allowing the ice sheets behind them to flow faster into the sea. Based on the correlation between global temperature and sea level in the geologic past, the potential for sea level rise from a business-as-usual 3°C global warming is easily tens of meters. Perhaps the sea level response will be slow enough not to matter in the coming century, but on the other hand there are times in the past when sea level rose by several meters in century timescales. The ice sheet models used to forecast sea level in the future are generally unable to explain the periods of rapid sea level rise in the past. All of this uncertainty in ice sheet response to climate makes it difficult to point to a definite CO_2 concentration that would be "safe" from provoking too much sea level rise.

A maximum temperature can be translated into a maximum CO_2 concentration in the atmosphere using a conversion factor called the climate sensitivity, ΔT_{2x}, defined as the eventual warming you'd get from doubling atmospheric CO_2. This is sort of a benchmark that climate modelers use to compare their models against each other, and the real climate. The value of ΔT_{2x} for the real world is based on analyses of the meteorological data from the last century, and from proxy data (tree rings and such) from the deeper past. The IPCC states a range of 2.5–4°C, with 95% confidence, for ΔT_{2x}. A good middle-of-the-road value is

3°C. The time period of maximum CO_2 concentration, the peak, will probably last a few centuries. Since it takes a few centuries for climate to fully respond to rising CO_2, the warming that could be expected within that time frame is probably a bit less than the full equilibrium value represented by ΔT_{2x}. Given the uncertainty in ΔT_{2x}, it has been calculated that an atmospheric CO_2 concentration of about 420 ppm would have a good chance of avoiding warming of 2°C

How much carbon is 420 ppm? The models don't all get the same answer, but they basically agree, in round numbers, that limiting atmospheric CO_2 to 420 ppm would require that the total emission, over the entire fossil fuel era, be limited to about 600 Gton C. Mankind has already released about 300 Gton C as CO_2 to the atmosphere. The additional 300 Gton C is comparable to the remaining reserves of oil and gas.

So one conceptually simple pathway to avoiding global warming would be to continue burning oil and gas, but just stop burning coal. Coal combustion is the source of only about a third of the carbon emissions today, with oil and gas each another third. In the long run, though, the amount of available coal exceeds oil and gas by a factor of ten or more. Ultimately, the future of Earth's climate comes down to decisions about coal.

Much of the political and scientific discussion of global warming is limited to the coming century, until the year 2100. Conveniently, by limiting the scope of consideration to 2100, and not caring about what happens afterward, there is a 67% bonus in the amount of CO_2 that can be emitted, raising the allowable emissions to about 1000 Gton C. The bonus comes from the expectation that the Earth takes several centuries to warm up completely after changing the atmospheric CO_2. Sixty-seven percent more CO_2 could be emitted before the year 2100, because in the year 2100, 67% of the warming from that CO_2 would not have happened yet. Global temperature could approach 2°C higher in the year 2100, but it could rise to maybe 3.3°C in the centuries thereafter. This strategy sounds a bit sneaky, but it is the perhaps unintended consequence of limiting the model runs, the warming scenarios and the political discussion to what happens by the year 2100.

For the coming decades, one simple way to look at the situation is that the natural world is taking up CO_2 about half as quickly as we are releasing it. Fossil fuel CO_2 is dissolving into the oceans, and being taken up by parts of the terrestrial biosphere (Chapter 8). If emissions were cut by about half, then the natural world would keep up with emission, and the CO_2 concentration in the atmosphere would stop rising, at least for a while.

In the larger scheme of things, having faith in human ingenuity, cutting by half doesn't seem unattainable. There are plenty of factors-of-two improvements that can be made to the way energy is used. Compact fluorescent light bulbs use a quarter of the energy of their incandescent cousins. Japanese hybrid automobiles are much more efficient than American SUVs. An average American emits twice the CO_2 as the average European or Japanese. Cutting U.S. emissions by half without sacrificing standard of living is demonstrably doable.

However, cutting U.S. emissions by half still wouldn't square us with the rest of the world. The average citizen of the United States, like the Europeans and Japanese, would still be emitting far more CO_2 than the average citizen of the Earth. The genuinely fair solution would be to divide the natural uptake rate of CO_2 today by the number of people on Earth. To reach that rate of carbon emission would require cuts in the developed world of about 80%. For the United States, Canada, and Australia, the cuts would be closer to 90%. As the ocean and the land biosphere saturate with the current higher concentration of CO_2 in the atmosphere, the natural uptake would slow, and the allowable emissions would decrease still further.

Realistically, drastic cuts of this magnitude are not going to happen overnight. Recognizing that atmospheric CO_2 concentrations are going to continue to rise for the time being, the IPCC has designed "stabilization scenarios," potential CO_2 trajectories for the future, which feature an eventual leveling out at some target concentration. The 450 ppm stabilization scenario reaches its target and levels out in sometime after about the year 2050 (Figure 22, top). Given the atmospheric CO_2 target, models of the carbon cycle and climate can be used to figure out how much CO_2 can be emitted while keeping to this CO_2 schedule. The emissions fluxes in Figure 22, bottom were generated using an on-line model that you can play with yourself at http://understandingtheforecast.com/Projects/isam.html.

While freezing CO_2 at its current concentration would require cutting emissions immediately by half, stabilizing CO_2 at 450 ppm would allow a few decades to go by while alternative energy sources are found. The business-as-usual alternative scenario has emissions doubling in the next 50 years from the present-day number of 7 Gton C / year to about 15 Gton C / year. Avoiding dangerous climate change requires emission cuts of around 80% by the year 2050.

CO_2 emissions originate from many different sectors of industry and the economy. To cut CO_2 emissions by a large fraction, changes must be made in many different CO_2 sources. The IPCC 2007 Mitigation Report calls the strategy a "portfolio of solutions" while most everyone else calls them "wedges." A wedge begins from zero and ramps up to a savings of 1 Gton C / year in the year 2050. Steve Pacala and Rob Socolow identified 15 potential wedges, all of which are based on technology and methods that already exist. Pacala and Socolow figured that seven wedges would be needed, but the danger limit CO_2 targets that people talk about have gotten lower in the last few years, so maybe 12 wedges could be necessary.

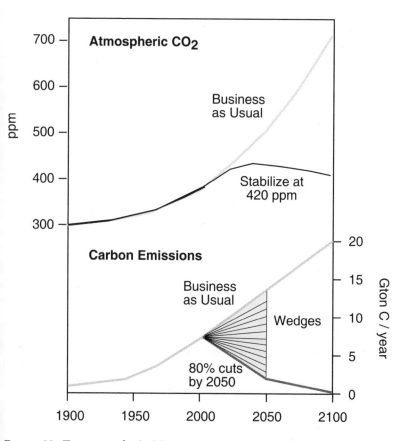

FIGURE 22. Top: atmospheric CO_2 concentration scenarios, business-as-usual and stabilization at 420 ppm. Bottom: CO_2 emission fluxes needed to achieve those CO_2 trajectories. The difference between business-as-usual and stabilization has been divided into "wedges."

One potential wedge is increased fuel efficiency for automobiles, from a business-as-usual 30 miles per gallon to a more efficient 60, assuming that in the year 2050 there will be 2 billion automobiles on the road, compared with 0.6 billion today. Another could come from changes in architecture such as insulation and lighting, which IPCC finds could be altered to save enough energy to produce a wedge, with overall savings of money. Fifty times more windmills than we have today would be a wedge. One sixth of the world's cropland devoted to bio-fuels would be a wedge. No-till agriculture could contribute a wedge.

Because there is so much coal on Earth, the climate of the future ultimately will be decided by what happens to the coal. A new technology for generating electricity from coal, called Integrated Gasification Combined Cycle (IGCC), is more expensive than just burning the stuff, so IGCC is not yet standard industry practice. In a regulated energy market, utility companies are often not even allowed to take any other than the cheapest course. An advantage of IGCC (among many, including efficiency) is that it produces CO_2 in a nearly pure form, whereas the effluent from a traditional coal-fired power plant is about 10% CO_2, diluted by nitrogen from the air. The CO_2 stream could be captured and injected back into the Earth in an approach called carbon capture and sequestration (CCS). When CCS is included as part of the process, it would be cheaper, over-all, to extract energy from coal using IGCC, rather than burning it and separating the CO_2 from the nitrogen in the effluent. IGCC also eliminates mercury and sulfur emissions.

There have been a few small efforts at CO_2 sequestration, mostly associated with the fossil fuel industry. CO_2 is separated from methane in the Norwegian Sleipner Vest gas project in the North Sea, and injected into a different rock layer than it was extracted from. CO_2 can be injected into depleted oil reservoirs through existing drill holes. The largest type of potential CO_2 repository in the Earth is called saline aquifers. These are veins of porous rock like sandstone, with salty water in the pores. The idea is that the water is salty, so no one would ever want it, so we might as well inject CO_2 into it. The IPCC Sequestration Report estimates that saline aquifers could hold up to 10,000 Gton of C as CO_2. Successful long-term gas sequestration in the

Earth is possible, as demonstrated by natural gas reservoirs that are millions of years old.

A leak of CO_2 from such a sequestration site could be deadly. When natural CO_2 escaped from Lake Nyos in Africa, the dense layer of odorless gas rolled across the landscape like a vengeful spirit, silently killing everything in its path. An earthquake could potentially release sequestered CO_2 from the Earth, or a break in a CO_2 pipeline aboveground. The easiest way to sequester CO_2 is to pump it into depleted oil and gas wells, since the holes have already been drilled. These have a potential for leaking through forgotten drill-holes, if not catastrophically like Lake Nyos, then perhaps more slowly, trickling out over the years.

I have seen a leakage rate of 0.1% per year proposed as a target for successful sequestration. At that rate, the CO_2 reservoir would escape to the atmosphere on a timescale of 1,000 years. Sequestration so defined would diminish the CO_2 peak but it would have little impact on the geologic-timescale impact of fossil energy on climate.

CO_2 could also be pumped into the deep ocean. The water near the injector nozzle would be fairly toxic to marine life, but the acidity would diminish as the CO_2 mixed into the rest of the ocean. Eventually, the CO_2 would begin to equilibrate with the atmosphere. After a millennium or so, the atmosphere would contain about a quarter of any CO_2 injected into the ocean, just as it would ultimately contain about a quarter of any carbon injected into the atmosphere. The final equilibrium state does not depend on where the CO_2 is released; it spreads out the same way in either case.

Ocean sequestration of carbon would reduce the amount of CO_2 in the atmosphere at its peak in the next couple of centuries. Releasing CO_2 into the atmosphere is arguably a worse idea than by-passing the atmosphere by injecting it into the ocean, if those were the only two options. But ocean sequestration would have minimal impact on the geologic-timescale impacts of fossil energy use.

After the year 2050, projections of energy use, compared with carbon cycle models of CO_2 stabilization, led Marty Hoffert and colleagues to conclude that some new sources of carbon-free energy are going to be needed. Large sources of energy. Their projections indicate that by the year 2100, humankind will require somewhere in the range of 10 to 20 terawatts of carbon-free power. The global total rate of energy production today, most of which is from fossil fuels, is 14 terawatts.

None of the carbon-free energy sources available today seem as if they could be ramped up to supply this much power. Using nuclear energy, for example, would require a new 1-megawatt power plant to be constructed every other day for the next 100 years to supply 10 terawatts. Biofuels would take up all of the agricultural land, leaving nothing for food. To do it using solar cells would require building them 10,000 times faster than we are doing today.

I have two personal favorite big ideas for generating lots of energy. One is to build solar cells on the moon, an idea advocated by David Chriswell at the University of Houston. There is no wind on the moon to cover solar cells with dust, no rain, no birds. No atmosphere and clouds would reflect the incoming sunlight away. The power could be beamed back to Earth as microwaves, a large beam that apparently wouldn't fry birds as they flew through. The energy from the beam would be received by an antenna on Earth maybe 10 kilometers on a side. Solar cells on the moon could be constructed from material refined from the lunar regolith, so the mass of the cells wouldn't have to be lifted up into space from the Earth's surface. It would take decades, technological developments (especially in robotics), and hundreds of astronaut tours of duty to construct this power source, but once construction got started, it could continue until it reached the required tens of terawatts of power.

My other favorite idea is high-altitude windmills, flying like kites in the jet stream. Electrical power can be transmitted through wires in the tether. For a nice artistic rendering of what this might look like, see windskypower.com. The power density (watts of energy per square meter of propeller area) is much higher at 30,000 feet elevation than it is down at the ground. High-altitude windmill power could also potentially scale up to generate the tens of terawatts of power we're looking for.

There are proposals to geo-engineer a return to a cooler climate. One is to blow smoke into the stratosphere. Sulfur could be added to airplane fuel, and it would produce a sulfur haze that would scatter incoming sunlight back to space. Suspended particles and droplets hang around a lot longer in the stratosphere than in the troposphere, because it never rains from the stratosphere. Volcanic eruptions such as Mt Pinatubo in 1992 or El Chichon in 1982 cool the Earth's climate for several years at a time by this mechanism. Another possibility would be to float a giant mirror in space, perhaps at the LaGrange point between the Earth and the Sun, where the gravitational pulls toward the two bodies are in balance with each other. The mirror could reflect sunlight away from the Earth, cooling it down.

These ideas seem a little thin when confronted with climate change that lasts for millennia. Aerosols in the stratosphere last for a few years and would require constant replenishment. Mirrors in space would probably require tweaking on timescales of decades or so; they wouldn't be stable in the right spot forever, I shouldn't think. If civilization in the future, facing some social collapse or economic depression, were unable to pay its climate bill (a bill that we left them, you're welcome, don't mention it), the climate impacts of all of our accumulated emissions would descend upon them all at once.

The only geo-engineering scheme that would address the issue squarely is to extract CO_2 from the atmosphere and sequester it someplace. The problem is that, in spite of all the fuss about CO_2 and global warming, CO_2 is only a trace component of the atmosphere, currently about 0.038% of the molecules in air. It takes energy and work to unmix CO_2 from that dilute mixture. Releasing CO_2 to the atmosphere only to extract it back out again would be a really stupid energy strategy. Of course trees manage to extract CO_2 from the atmosphere, and farm crops. Farm waste could be dumped into the deep ocean as a carbon sequestration measure. But all of this takes work; it appears that it would be more sensible to not release the CO_2 to the air in the first place.

The merits of various options for dealing with the issue of climate change are sometimes compared within the framework of economics. Changes in the infrastructure of global energy production are likely to entail some cost. Energy efficiency probably requires an investment up front, but might pay for itself in the long run. Damages from climate change might also be assessed in terms of financial cost, as well as the effort required to adapt to climate change. In some particulars, there may be benefits to a change in climate, which might be valued monetarily. The tools of economics might allow us to search for the pathway of least cost.

First the disclaimer. Economic models are even dodgier than climate models. This is not because climate scientists are smarter than economists, but rather because economic forecasting is just plain hard. Economies are driven by technological progress, predictions of which are usually laughable in hindsight. But for what it's worth, economic projections of the costs of avoiding dangerous climate change are typically a few percent of global economic production per year.

This sounds like a lot of money, and if it were heaped into a pile of bills in the middle of Central Park, it would certainly be a very large pile. But economies tend to grow. Let's say that the cost of developing and using only clean energy amounts to 3% of GDP every year. Let's also say that the economy is growing at a rate of 3% per year. Wait a year, and the GDP will grow to the size of the dirty-energy economy, just one year later. If the rate of growth were 1.5% per year, the clean economy would lag two years behind the dirty one.

The costs of beginning to lower CO_2 emissions are generally not too high. According to the Mitigation report of the IPCC, the building sector of the economy, in particular, could cut emissions substantially at net negative cost (that is, saving money by saving energy). If emission cuts are inevitable, it makes economic sense to be ahead of the curve, developing efficient technologies now to sell to others, rather than indulging in delay, ceding this business opportunity to others (like European windmill manufacturers and Japanese makers of efficient automobiles). Delaying CO_2 cuts would mean that the eventual cuts would have to be deeper in order to reach the same stabilized atmospheric CO_2 concentration. For this reason also, if cuts are inevitable, it would be cheaper to start now rather than to dither.

The free market has a blind spot, apropos to the problem of CO_2 emissions, called the tragedy of the commons. As we've discussed earlier, the situation arises when many people share the benefits of a common resource. The classical example is a common field for grazing livestock, but the chemistry of the atmosphere is another example of a shared commons. Benefits can be extracted from the commons by individuals, either for feeding sheep or for dumping CO_2, but the costs of using the commons are paid by everyone. Economists call the degradation of the commons an external cost, because it is external to the budgets of the individual decision-makers. The end result is that the common resource gets over-exploited, because it is in the interest of each individual to grab as much as he or she can.

In the language of economics, the way to avoid distorting the market (that is, doing everyone harm) is to internalize the cost of climate change (that is, make the person buying the gasoline pay for the climate change). Some form of regulation is required, either a form of tax to internalize the cost, or a restriction on emissions, to avoid the tragedy of the commons.

Economics has the property of being rather myopic over the course of time. The cause of the short-sightedness is the interest rate. When money is invested, it grows by accruing interest. $100 invested today will grow to, say, $103 at the end of the year, and $2000 in a century, if the interest rate is 3%. Imagine an economist faced with a cost in the future, and a way to fix the problem by paying now. Should she pay now, or pay later? If she pays later, she can invest a smaller sum of money now, and it will grow to meet the obligation later. After 100 years, a cost of $100 shrinks to just $5 today. It's much cheaper to pay later. An interest rate of 3% tends to limit our perceptions to a time horizon of about 30 years. The tool of economics is simply not programmed to pay attention to things that are too far in the future.

Economics is more than just a set of tools, however. It is a description of the way that money, the lifeblood of our economic system, really flows, analogous to the statement that water flows downhill. Money flows toward short-term gain, and toward over-exploitation of unregulated common resources. These tendencies are like the invisible hand of fate, guiding the hero in a Greek tragedy toward his inevitable doom. Our understanding of economics tells us that the free hand of the market, also known as business-as-usual, will not cope gracefully with the threat of global warming.

Ultimately the question may come down to ethics, rather than economics. Slavery, only abolished a little over a century ago in the United States, was an ethical issue. Ultimately it didn't matter whether it was economically beneficial or costly to give up. It was simply wrong.

The costs and benefits of fossil fuel use are not shared fairly. At the present day, the benefits of the fossil fuel economy accrue mostly to the industrialized nations in the temperate latitudes, while the IPCC Working Group II Scientific Assessment Report on Impacts of Climate Change finds that the costs of climate change will be paid most dearly in the tropics. More people are killed when any sort of natural disaster strikes in the developing world than in an industrialized nation, where emergency services are better equipped. Low-latitude countries tend to have a higher proportion of subsistence farmers, and therefore greater immediate vulnerability to changes in the weather. In the industrialized world, energy-intensive farming methods are producing food faster than we can eat it. Food production is also more globalized in the developed world, insulating people somewhat from dependence on local agricultural conditions.

There is also a divide in time between the winners and the losers. The benefits to using fossil fuels accrue now and into the coming century until the fuel runs out, while the costs will last for millennia. Most of the people impacted by global warming, numerically, are people in the future. Earthlings a century from now do not even have an economic vote in how we conduct our affairs; their right to vote has been discounted to nothing by the economic interest rate.

Ethics and fairness are a lot to ask of the political process, especially when most of the people affected by the decision, people of the distant future, do not have a voice in the decision.

We will conclude by considering the awesome potential energy impacts of a gallon of gasoline on Earth. When it is burned, it yields about 2500 kilocalories of energy, but this is just the beginning. Its carbon is released as CO_2 to the atmosphere, trapping Earth's radiant energy by absorbing infrared radiation.

About three-quarters of the CO_2 will go away in a few centuries, but the rest will remain in the atmosphere for thousands of years.

If we add up the total amount of energy trapped by the CO_2 from the gallon of gas over its atmospheric lifetime, we find that our gallon of gasoline ultimately traps one hundred billion (100,000,000,000) kilocalories of useless and unwanted greenhouse heat. The bad energy from burning that gallon ultimately outweighs the good energy by a factor of about 40 million.

The enormous world-altering potential of that gallon of gasoline has taken the reins of Earth's climate away from its natural stabilizing feedback systems, and given them to us. May we use our newfound powers wisely.

Further Reading

Chapter 1

Spencer Weart, *The Discovery of Global Warming*, 2003.
David Archer, *Global Warming: Understanding the Forecast*, 2006.
IPCC Scientific Assessment, Volume 1, Chapter 1, Historical Overview of Climate Science, 2007.

Chapter 2

IPCC Scientific Assessment, Working Group I, The Scientific Basis. Summary for Policymakers and Technical Summary, 2007.
John Houghton, *Global Warming: The Complete Briefing*, third edition, 2004.

Chapter 3

Elizabeth Kolbert, *Field Notes from a Catastrophe*, 2006.
Mark Lynas, *High Tide: News from a Global Warming World*, 2004.
IPCC Scientific Assessment, Volume 1, Summary for Policymakers, Technical Summary, and Chapters 10, Global Climate Projections, and 11, Regional Climate Projections, 2007.

Chapter 4

Brian Fagan, *The Long Summer*, 2004.
Brian Fagan, *The Little Ice Age: How Climate Made History 1300–1850*, 2000.
Committee on Abrupt Climate Change, National Academy of Sciences, *Abrupt Climate Change: Inevitable Surprises*, 2002.
IPCC Scientific Assessment, Chapter 7, Paleoclimate, 2007.

Chapter 5

John Imbrie and Katherine Palmer Imbrie, *Ice Ages: Solving the Mystery*, 1979.

Chapter 6

David Archer, *Global Warming, Understanding the Forecast*, Chapter 7, Carbon on Earth, 2006.
Robert Berner, *The Phanerozoic Carbon Cycle: CO_2 and O_2*, 2004.
Lee R. Kump, James F. Kastin, and Robert G. Crane, *The Earth System*, 2004.

Chapter 7.

Brian Fagan, *The Long Summer*, 2004.
Brian Fagan, *The Little Ice Age: How Climate Made History 1300–1850*, 2000.
Jared Diamond, *Collapse: How Societies Choose to Fail or Succeed*, 2005.
Owen B. Toon and colleagues, Consequences of Regional-Scale Nuclear Conflicts, *Science*, 315, 1224–1225, 2007.

Chapter 8

Wally Broecker and Taro Takahashi, Neutralization of fossil fuel CO_2 by the oceans, in *The Fate of Fossil Fuel CO_2 in the Oceans*, edited by Anderson and Malahoff, 1978.
David Archer, Fate of fossil fuel CO_2 in the geologic time, *Journal of Geophysical Research Oceans*, doi:10.1029/2004JC002625, 2005.
David Archer and Victor Brovkin, Millennial atmospheric lifetime of anthropogenic CO_2, *Climatic Change*, 2007, in press.
Timothy Lenton, Enhanced carbonate and silicate weathering accelerates recovery from fossil fuel CO_2 perturbations, *Global Biogeochemical Cycles*, 20, doi:10.1029/2005GB002678, 2006.
Phillip Goodwin and colleagues, Ocean–atmosphere partitioning of anthropogenic carbon dioxide on centennial timescales, *Global Biogeochemical Cycles*, 21, doi:10.1029/2006GB002810, 2007.
Andrew Ridgwell and J. Hargreaves, Regulation of atmospheric CO_2 by deep-sea sediments in an Earth system model, *Global Biogeochemical Cycles*, 21, doi:10.1029/2006GB002764, 2007.

Chapter 9

Royal Society of London, *Ocean Acidification Due to Increasing Atmospheric Carbon Dioxide*, 2005.

Toby Tyrrell and colleagues, The long-term legacy of fossil fuels, *Tellus* B, 59, 664, 2007.

Chapter 10

Marten Scheffer and colleagues, Positive feedback between global warming and atmospheric CO_2 concentration inferred from past climate change, *Geophysical Research Letters*, 33, doi:10.1029/2005GL025044, 2006.

David Archer, Methane hydrate stability and anthropogenic climate change, *Biogeosciences*, 4, 521–544, 2007.

ACIA contributors, *Arctic Climate Impact Assessment, Scientific Report*, 2005.

Chapter 11

Michael Oppenheimer, Global warming and the stability of the West Antarctic ice sheet, *Nature*, 393, 325–334, 1998.

Richard Alley and colleagues, Ice sheet and sea level changes, *Science*, 310, 456–460, 2005.

H. Jay Zwalley and colleagues, Surface melt-induced acceleration of Greenland ice-sheet flow, *Science*, 297, 218–222, 2002.

Goran Ekstrom and colleagues, Seasonality and increasing frequency of Greenland glacial earthquakes, *Science*, 311, 1756–1760, 2006.

Jim Hansen, Scientific reticence and sea level rise, *Environmental Research Letters*, 2, 024002, 2007.

Jim Hansen, A slippery slope: How much global warming constitutes "Dangerous Anthropogenic Interference"? *Climatic Change*, 65, 269–279, 2005.

Chapter 12

David Archer and Andrey Ganopolski, A movable trigger: Fossil fuel CO_2 and the onset of the next glaciation, *Geochemistry, Geophysics, Geosystems*, 6, doi:10.1029/2004GC000891, 2005.

Didier Paillard, Glacial cycles: Toward a new paradigm, *Reviews of Geophysics*, 39, 325–346, 2001.

Epilogue

Steve Pacala and Rob Socolow, Stabilization wedges: Solving the climate problem for the next 50 years with current technologies, *Science*, 305, 968–973, 2004.

Marty Hoffert and colleagues, Energy implications of future stabilization of atmospheric CO_2 content, *Nature*, 395, 881–885, 1998, and *Science*, 298: 981–988, 2002.

David Criswell, Solar power via the Moon, *The Industrial Physicist*, Apr/May 2002.

Index